Studies in Systems, Decision and Control

Volume 73

The series "Studies in Systems, Decision and Control" (SSDC) covers both new developments and advances, as well as the state of the art, in the various areas of broadly perceived systems, decision making and control–quickly, up to date and with a high quality. The intent is to cover the theory, applications, and perspectives on the state of the art and future developments relevant to systems, decision making, control, complex processes and related areas, as embedded in the fields of engineering, computer science, physics, economics, social and life sciences, as well as the paradigms and methodologies behind them. The series contains monographs, textbooks, lecture notes and edited volumes in systems, decision making and control spanning the areas of Cyber-Physical Systems, Autonomous Systems, Sensor Networks, Control Systems, Energy Systems, Automotive Systems, Biological Systems, Vehicular Networking and Connected Vehicles, Aerospace Systems, Automation, Manufacturing, Smart Grids, Nonlinear Systems, Power Systems, Robotics, Social Systems, Economic Systems and other. Of particular value to both the contributors and the readership are the short publication timeframe and the world-wide distribution and exposure which enable both a wide and rapid dissemination of research output.

Indexed by SCOPUS, DBLP, WTI Frankfurt eG, zbMATH, SCImago.

All books published in the series are submitted for consideration in Web of Science.

Octavian Iordache

General Reference Architecture Frameworks

Springer

Octavian Iordache
Dollard-Des-Ormeaux, QC, Canada

ISSN 2198-4182 ISSN 2198-4190 (electronic)
Studies in Systems, Decision and Control
ISBN 978-3-031-70717-9 ISBN 978-3-031-70718-6 (eBook)
https://doi.org/10.1007/978-3-031-70718-6

This Springer imprint is published by the registered company Springer Nature Switzerland AG
The registered company address is: Gewerbestrasse 11, 6330 Cham, Switzerland

For it is the same thing that can be thought and that can be.
Parmenides

… that there is in the being of things something which corresponds to the process of reasoning, that the world lives, and moves, and has its being, in a logic of events.
C. S. Peirce, 1898

The 'scientific part' of chemical engineering consists in breaking down real complex systems into subsystems, which are then described using our understanding of fundamental chemical and physical processes. The 'engineering part' of chemical engineering consists in using this new-found knowledge in the design and construction of a working plant which is capable of producing the desired product, even if our understanding of the single subsystems is today incomplete.
K. Wintermantel, 1999

Preface

Studying high complexity in engineering and science, in reality and virtuality is the object of this book. Complex systems are assemblies of several systems characterized by emergent behavior resulting from nonlinear interactions. The main source of complexity is the multiplicity of processes, scales, fields, dimensions, disciplines, and logic rules for the considered systems. Multiple interactions and emergences are the core of higher complexity and of associated models and methods for projects implementations.

The starting point of our approach is the observed similarity or isomorphism of roadmaps towards higher complexity and of reference architectures for different domains.

Logical reference architecture of our virtual conceptual schemes corresponds to the structural reference architecture of reality. Systems ranging from inorganic to biological, cognitive, intelligent, to design and modeling mathematical models share similar roadmaps and similar reference architecture frameworks.

The objective is to propose General Polytopic Roadmaps (GPTR) and General Reference Architecture Frameworks (GRAF) and to use these for 8D Program implementation.

The GPTR shows the stages: 0D, 1D, 2D, 4D, and 8D. They correspond, respectively, to:

Descriptive, Adaptive, Evolvable, Self-Evolvable (SE), and SE of SE systems.

For example, correlating the real physical with virtual cyber in cyber-physical systems, allows developing Industry 4.0 or in other words, 4D, live-like or SE systems. This is the A-Life or Industry 4.0 evolution stage. Then, by grafting 4D for any component of the existing 4D live-like systems, the 8D intelligent-like, or SE of SE systems will result.

This is the A-Intelligence or Industry 8.0 evolution stage. The 4D stage is devoted to replace physical activities, while the 8D stage tries to replace mental activities of humans.

The GPTR and the GRAF are committed to explore the high complexity in different domains. High-dimensional reference architectures should be envisaged in industry giving that the conventional operations, equipments, methodologies, or organizations exploring the ever-growing complexification reached their limits and need both 4D, live-like, as 8D, intelligent-like capabilities. The exploration of ever growing complexities needs the high dimensional polytopic projects implementation. Polytopic projects impose connecting science and engineering, virtuality and reality.

The book is divided into eight chapters. Chapter 1 introduces the GRAF. This is presented as a 4D hypercube of 4D hypercubes that is as an 8D polytope. Conservative and innovative strategies of evolution, from 4D to 8D, are presented.

Chapter 2 emphasizes the role for the dialogue of processes in duality, of the logic of contradiction, of iteration, and of included middle to explore high complexity. The role of a general method and of included middle or Systems 3, between System 1 and System 2 frames of dual process theory, is revealed. Chapter 3 refers to operations and equipments of chemical engineering interest as permutations, mixings, and separations. Chapter 4 refers to modeling. Here the road to complexity in modeling, to digital twin, and to digital twin of digital twin is illustrated. Model-Based Engineering case studies are analyzed. Chapter 5 concerns creative design models. Dual process design, processes integration, divergence, and convergence design models are presented. Industry 4.0, future developments to Industry 8.0, and chemical engineering paradigms are examined in Chap. 6.

Chapter 7 focuses on complex systems study. Production systems as systems of systems architecture frameworks, decision models, operations process, and cyber-physical social systems have been presented in the new general frame.

Chapter 8 analyzes implementation of high-complexity projects for different levels of reality. The unifying and diversifying strategy for reality and virtuality, and the 8D Program Manifesto closes the Chap. 8.

Despite the fact that the majority of case studies are related to chemical engineering the presented methodologies aim to be universal and equally applicable to other fields of engineering or science. The book will be useful to engineers, researchers, entrepreneurs, and students in different branches of production, science, and engineering of complexity.

Montreal, Canada Octavian Iordache
June 2024

Contents

Abbreviations

CPS	Cyber-Physical Systems
DoE	Design of Experiments
DPT	Dual Process Theory
DT	Digital Twin
GPTR	General Polytopic Roadmaps
GRAF	General Reference Architecture Frameworks
IM	Included Middle
MBE	Model-Based Engineering
MBIC	Material-Biologic-Intelligent-Cognitive
OODA	Observe Orient Decide Act
PTP	Polytopic Project
PTR	Polytopic Roadmap
RAMI	Reference Architecture Model for Industry
SE	Self-Evolvable
SO	Self-Organized
SoS	System of Systems

List of Figures

List of Tables

Chapter 1
Complexity

Abstract A way to explore complexity and emergence is presented here. Reference architectures are introduced as a set of schemes that identifies structures and allows integration of assets, operations, equipments, methods, products, tests and applications in different projects. The proposed main stages for high complexity roadmap are: Descriptive, Adaptive, Evolvable, Self-Evolvable (SE) and SE of SE. They correspond to 0D, 1D, 2D, 4D, and 8D reference architectures. The SE stage is associated to 4D and live-like systems. The SE of SE stage is associated to 8D and to intelligent-like capabilities. General roadmaps and frameworks for 4D and 8D reference architectures are introduced.

1.1 Complexity and Emergence

What is a complex system, and what does it means for a system to show high complexity and emergence?

The study of complex systems or more generally the science of complexity has been a hot research topic for the last decades.

A complex system is described as a structure or process involving non-linear interactions among many parts and levels, which displays emergent properties. In other words this means that the aggregate system activity is not derivable from the linear summations of the activity of individual components and that novel structures, patterns or properties arise, from interactions among parts.

Complex systems are ones in which patterns can be seen and understood, but interplay of individual parts cannot be reduced to the study of individual parts considered in isolation from one another. A survey of the literature indicates that there is no standard agreed upon definition of a complex or emergent system. Some of the existing definitions may even seem contradictory but they may make sense when applied to specific types of systems and from which perspective one choices to observe (Adami 2002; Kauffman 1995). This suggests considering several domains of complexity and a hierarchy of levels for complexity. The complexity for inorganic systems differs from the complexity for biological, cognitive or intelligent systems.

O. Iordache, *General Reference Architecture Frameworks*, Studies in Systems, Decision and Control 73, https://doi.org/10.1007/978-3-031-70718-6_1

An example of physical complex system is the global climate, including all components of the atmosphere and oceans and taking into account the effects of extraterrestrial processes as solar radiation and meteorites. An illustration of complex biological system is the human brain composed of millions of nerve cells. Their collective interaction allows recognizing visual, acoustic or olfactory patterns, speaking and performing different mental activities.

An example of complex social system is the human society with its participants, natural resources and capital goods, financial and political systems. For the logical and mathematical realm, examples of high complex calculus systems may consists of large scale distributed software systems or hierarchies of layered computing subsystems self-organized and running together to achieve particular objectives.

What is remarkable is that systems that have apparently little in common-material systems as an array of polymers in a test tube, biological systems as a group of receptors on a cell's surface, knowledge or cognitive systems as a group of ants in a swarm or human agents in a company-often share remarkably similar structures and means of organization. This explains and justifies the need for a science of complexity.

Features such as non-linearity, hierarchy of levels, time-scales, connectivity, non-equilibrium, unpredictability, interconnectivity, collective behavior, self-evolvability, self-organization, self-production, self-reference, and multiple agencies are associated with complexity studies. Complexity is correlated to non-linearity, which is a necessary but not sufficient condition of complexity, as well as to interconnectivity, self-organization, self-evolvability, self-similarity and collective behavior (Mainzer 1996).

The understanding of complexity changes with the domains of application. Surveys consider that complexity has not an absolute meaning, and it is only a relative notion depending on the level of observation or abstraction. It is commonly stated and accepted that some objects and processes are more complex than others.

We must to take into account this facet of complexity as a relative concept which depends both on the task at hand and on the tools available to achieve this task.

For industrial systems, despite the fact that numerous physical, chemical or biological processes are identified as complex, some of the conventional ones may be operated in regimes were complexity properties are neglected. For several centuries, physical and chemical sciences made great steps by experimenting and constructing simplified models of complex phenomena, deriving properties from the models, and verifying those properties by new experiments. This approach worked because the complexities ignored in that models were not the critical properties of the phenomena. It does not work when the complexity becomes the essential characteristic. In a continuously increasing number of cases the complexity is not transient or atypical, but it is an intrinsic, basic property of that systems. Given this situation the challenge for engineers, scientists and entrepreneurs is not only to identify complexity domains but also to show how to overtake the successive complexity barriers. The next defy in science, technology and economy is to explore the complexity, finding the ways from high complexity towards a new simplicity. The 21st century problems as energy, nutrition, health, ecology, finance and security pertain without doubt to high complexity sphere and need new methodologies.

As for complexity, there is not a consensus on a standard definition of emergence However, emergence is the property that distinguishes and makes complex systems mandatory to study. Emergence is the phenomenon that differentiates complex systems from complicated or multi-component ones. A complicated device is a simple addition of separable preexistent components, whereas a complex system is made out of a large number of components interconnected by a network of relations in such a way that a modification somewhere in the system modifies the whole system.

Emergence is what parts of a system do together that they would not do by themselves. An example is the collective behavior, that a system does by virtue of its relationship to its environment but it would not do by itself. Important facets of emergence concept are: novelty, creativity, micro–macro effects, coherence, nonlinearity, interacting and segregating parts, local and global control, robustness, and flexibility.

A class of definitions for emergence emphasizes the role of multiple levels of organization and their interaction. The starting point is that the large majority of complex systems exhibit hierarchical self-organization in levels under selective constraints.

Self-organization will occur when individual independent parts in a complex system interact in a jointly cooperative manner that is also individually appropriate, such as to generate a higher level organization.

Complex systems can be observed at different levels of investigation. For example we can observe a chemical industrial installation at the level of molecules, at the level of devices interactions, or at the environmental level.

A property may be considered as emergent if it cannot be explained from the properties and interactions of the lower level entities. Obviously, it is mandatory to explain from were novelty comes. This kind of approach requires clarifying the relation between higher levels and lower levels in hierarchy. Such study refers to the micro–macro, local to global or the two-ways mechanism of emergences and also to the concept of downward causation. An emergent feature is supposed to have some kind of causal power on lower level entities. Downward form of causality, operates from upper to lower level, and complements the upward causation. Lower level entities exercise an upward causation on the emergent features.

This approach implies a two-levels and two-way causal relation. As an illustrative example we may consider several component transport processes organizing in a chemical engineering installation in evolution. The component transport processes affect how the installation develops, upward, and the development of the installation affect the behavior and the interaction of the component processes, downward.

As a working approach, the emergence is considered as a dynamic, nonlinear process that leads to novel properties, structures, patterns at the macro-level or global level of a system from the interaction of the part at the micro-level or local level. Such novelty cannot be understood by reductionism but may be studied by looking at each of the parts in the context of the system as a whole. Detecting and breaking complexity frontier allows in fact exploiting properties of emergence, as non-linearity of interactions and coherence. This may be accomplished by distributed systems constructed as a group of interacting autonomous entities that are designed to cooperate, to have an emergent globally coherent behavior. Due to non-linear reinforcement, local interactions may result in a larger effect in the form of a novelty at the macro-levels.

1.2 Reference Architecture

Reference architectures are standardized architectures that provide a frame of reference for a domain. Reference models or architectures provide a common vocabulary, reusable designs, and industry best practices. They are not solution architectures, that is, they are not implemented directly. Rather, they are used as a constraint for more concrete architectures. Reference architectures include common architecture principles, patterns, building blocks, and standards.

Reference architecture may be a schematic or set of schematics a document or set of documents that identify recommended structures and integration of assets, methods, products and services in a project to come up with an appropriate solution or set of solutions to problems (Muller 2012).

The reference architecture embodies accepted industry or society best practices, noting criteria specific to the project being developed, as well as listing required functions, recommended processes to follow, and suggesting the delivery method for specific domains.

The ultimate purpose of reference architecture is to provide developers, architects, operators, engineers, designers and entrepreneurs with guidelines, to keep in mind and conserve as they learn, research, plan, develop, and produce a new product, or new project without being so restrictive that it becomes an obstacle to innovations.

One goal of the reference architecture is to offer enough core prerequisites that the developer can avoid wasting time on details and focus on the new solutions and frames.

Reference architectures provide a frame of reference that helps to get an overview of a domain and they provide a starting point for new product architecture effort.

They provide entrepreneurs with basic structures.

The process of creating a reference architecture is a way for a consumer to clarify and check the details before it goes to a developer, which helps shorten the development time, contains costs, and can serve as a check-off sheet to keep everyone involved on the same page throughout the project.

Developers use reference architectures to clarify necessary development processes, avoid roadblocks, keep teams on track, avoid cost overruns, test and verify final results.

There are modeling reference architectures, software reference architectures, banking or building reference architectures, automotive reference architectures, and so on.

For instance, the reference architecture in the field of software architecture provides a template solution for architecture for a domain. It also provides a common vocabulary with which to discuss implementations, with the aim to stress commonality.

The software reference architecture is a software architecture where the structures and respective elements and relations provide templates for concrete architectures in a domain.

Reference architectures reduce costs and cut development times of new products, for all parties involved. The reference architecture captures the essence of the architecture of a collection of systems, that is, it needs systems, systems of systems, and so on.

The purpose of reference architectures is to provide guidance for the development of architectures for new versions of the system or extended systems and product families. These provide guidelines for the content of the reference architecture and the process to create and maintain it.

The reference architecture is created by capturing the essentials of existing architectures and by taking into account future needs and opportunities, ranging from specific technologies and equipments, to patterns to business models and market segments.

For numerous domains we can see two coincident trends:

- Increasing complexity, scope and size of the system of interest, its context and the organizations creating the system.
- Increasing dynamics and integration: shorter time to market, more interoperability, rapid changes and adaptations in the field, in a competitive market, with cost and performance pressure.

These trends cause a transition from simple closed system creation to complex distributed open system evolution. In the simple and closed situation, a system could be created at one location, by one supplier, in one organizational entity.

Many of today's systems are developed as complex distributed open development at multi-locations, by multi-suppliers, across multi-organizations.

To account for multiplicity we make use of the label polytopic. The multiplicity is not limited to organizations, suppliers and locations. Systems also outline multi-discipline, multi-application, multi-cultural features, while development and manufacturing is based often on multi-sourcing, multi-logics and so on. Polytopic reference architectures start to appear in organizations where the multiplicity and the resulting complexity reaches a critical mass triggering a need to facilitate product development and life-cycle support in the complex distributed world. The high complexity and high dimensionality requires polytopic projects and Polytopic Roadmaps, (Coxeter 1973; Iordache 2012).

Coxeter defined polytope as the general term of the sequence: point, segment, polygon, polyhedron, and so on.

This justifies the term polytopic associated to the projects, frameworks, and roadmaps considered here.

The reference architecture provides:

- Common lexicon and taxonomy, for example by a domain model
- Common architectural vision
- Modularization and the complementary context

The common lexicon and taxonomy facilitates communication across the multiple dimensions. The common architectural vision focuses and aligns efforts of multiple peoples and teams. Modularization helps to divide the effort, where the context information ensures later integration (Muller 2012).

Architecture frameworks provide approaches to describe and identify necessary inputs to a particular architecture as well as means to describe that architecture. So, architecture frameworks give enterprise architects the tools they need to adequately describe and collect requirements, without mandating any specific architecture type.

Architecture frameworks describe the taxonomy of the kinds of architectural views that an architect might consider developing and provides guidelines for making the choice for developing particular views.

Reference architecture goes one step further by accelerating the process for a particular architecture type, helping to identify which architectural approaches will satisfy particular requirements, and figuring out what a minimally acceptable set of architectural artifacts are needed to meet the best practices requirements for a particular architecture. Reference architectures emphasize the template part of the definition of architecture. Both frameworks and reference architectures provide best practices, and while it might be argued that reference architectures provide more of a methodology than a framework does, reference architectures are still not really characterized by their methodology component. Most can be characterized by their template component. From this perspective, patterns are instances of templates in the context. In fact, multiple reference architectures for the same domain are allowable and useful.

1.3 Roadmaps

Several scientific and engineering studies show that evolvability observed in natural systems is based on similar sequences of reference architectures as the evolvability of concepts in human mind and of associated scientific and engineering projects. Such studies support the idea of parallelism between virtuality and reality or between science and engineering.

As an example of roadmap for complexity we may consider the roadmaps of numbers as shown in Fig. 1.1.

It refers also to the normed division algebras (Schafer 1966; Springer and Veldkamp 2000). There exist exactly 4 objects which qualify as normed division algebras: R the one dimensional Reals, C the two-dimensional Complexes, H the four-dimensional Quaternions, and O the eight-dimensional Octonions.

The only normed division algebras, which have a norm satisfying $|zw| = |z||w|$ have dimension 1, 2, 4, or 8.

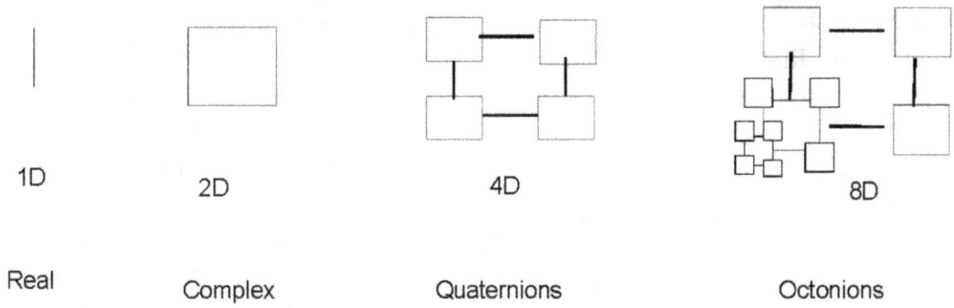

Fig. 1.1 Roadmap for numbers

The real numbers are represented in Fig. 1.1 as a 1D line. Complex numbers needs two axes that is a 2D plan. They are represented as 1D of 1D. Quaternions are 4D and are represented here as 2D of 2D. The dimension is 4D = 2D + 2D. Octonions result by iterating the procedure to obtain quaternions and are represented here as 4D of 4D.

The 8D is partially represented in Fig. 1.1 to avoid complicated drawings.

In the 8D frame each step of 4D is replaced by 4 component sub-steps. These sub-steps have 4 sub-sub-steps. In this way a 4D of 4D frame is resulting.

The dimension is 8D = 4D + 4D.

The doubling of dimensions from the real numbers R to the complex numbers C to the quaternions H can be taken as one stage further, to the octonions O.

By increasing dimensions the algebra lose axioms. From R to C, we lose the ordering, from C to H, we lose commutativity while from H to O, we lose associativity.

By doubling the dimension again, we lose the multiplicative property of the norms.

As another example of mathematical roadmap we refer to the Leibniz-Loday algebras: associative algebras, dialgebras, quadri-algebras, and octo-algebras (Leroux 2003; Ebrahimi-Fard and Guo 2005; Loday and Vallette 2012). The Leibniz-Loday algebras form the roadmap of algebras with a common property of splitting associativity.

It was observed that systems as industrial equipments, telecommunication networks, and socio-technical systems as manufactures, companies and societies evolve and became more and more complex. It is the result of natural development for products and markets, of the embedding of hardware, software, logical or mathematical tools and necessary knowledge into such systems.

The ever-growing complexity of modern industrial and economic systems led to unsustainable increases in their management and operation costs. Problems of design and organization become pressing as products increase in complexity regarding modeling, hardware and software. It is unavoidable to shift much of the organization tasks into the machines and installations themselves. This brings up the problem of keeping the self-organization under control. Product, installations, hardware and software engineers aim to relieve growing complexity problems by conceiving learning, evolvable, self-evolvable (SE) that is, life-like systems and SE of SE that is, intelligent-like systems.

Live-like and intelligent-like systems should be able to adapt dynamically to the current conditions and to explore future requests of the environment.

Intelligent policies place emphasis on the future effect of any given policy or business practice on humans, economy, and ecology.

Conventional engineering or scientific methods have been conceived to deal with systems reducible to simpler parts which exhibit controlled behavior. A significant trend in high complexity managing was and continues to be the pursuit of autonomy for systems. However, the autonomous robotic systems will inevitably fail to meet some tasks or obligations when operated in unexpected field conditions, if they are not designed for live-like and intelligent-like capability (Huang et al. 2004).

Live-like and intelligent-like behaviors have been proposed as the key requirements to explorer ever-growing complexity and to successfully operate in high complexity conditions. The associated capabilities are the paradigms of artificial-life and artificial-intelligence proposed to respond to the shift to faster and stronger hardware integrated computer systems, to the need for self-management of high complexity systems.

The live-like capability replaces physical activities while the intelligent-like capability replaces mental activities of people.

We start from the observation that there are significant differences between adaptable, evolvable, self-evolvable (SE) and SE of SE systems as stages of increasing autonomy. Gradually it will be possible to make a system more close to live-like and to intelligent-like ones as the complexity understanding and resources to prevail over high complexity will grow (Huang et al. 2004).

Insights for the high dimensional reference architectures are offered by different scientific domains. Gardiner et al. (2010) use EEG data as evidence of higher dimensional fractal signals in the brain, ranging fractionally from 5 to 8D. EEG data is a good place to look for the underlying patterns of chaos in the brain since it samples many millions of neurons simultaneously. Recent studies have arrived at a dimension between 5 and 8D for human EEG data. This suggests that the brain operates in a higher than 4D dimension of traditional space–time of physics.

High dimensionality implications for brain and artificial intelligence have been examined by Reimann et al. (2017, 2022). A group of four neurons has dimension three, five neurons dimension four and so on. The more neurons in a group, the higher are the dimension-and so the system becomes increasingly complex. In such studies, dimension does not describe spatial dimensions, but rather the topological dimension of the geometric objects we are describing. The observed 7D or 8D simplexes are still considered as included in the physical 3D space.

High dimensionality was also encountered in data processing and neurodynamics (Cowan 2000). According to Cowan the capacity of short-term memory is limited to the number of four items to which attention can be simultaneously directed. There exists a central capacity limit of four chunks in short-term explicit memory presumably corresponding to the focus of attention. This theory assumes that attention is depending on oscillation of cortical potentials.

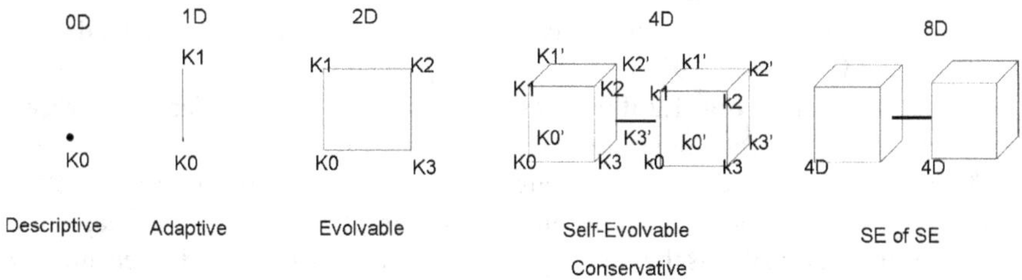

Fig. 1.2 Polytopic roadmap for complexity-conservative strategy

Four or eight levels are necessary for data analysis and interoperability (Rossiter and Heather 2003). Less than four offers only local interoperability. More than four or eight should be formally defined and may offer benefits for interoperability.

The polytopic architecture is illustrated by the evolution of reference architectures for high complexity as shown in Fig. 1.2. Figure 1.2 shows the succession of stages in understanding, modeling and implementation of systems as complexity grows (Iordache 2012, 2013, 2018). It is a road that allows facing high complexity advent.

Here K0 denotes states or the level 0 of conditioning. K1, K2, K3 denotes conditions, meta-conditions and meta-meta conditions. Minuscule.k0, k1, k2, k3 denote the associated control.

Figure 1.2 shows the Polytopic Roadmap for complexity-conservative strategy.

The successive stages of the complexity sequence are represented as dots (0D), segments (1D), squares (2D), hypercubes (4D) and hypercubes of hypercubes (8D).

The dimensions of stages shown in Fig. 1.2 are:

- 0D-Descriptive
- 1D-Adaptive
- 2D-Evolvable
- 4D-Self-Evolvable
- 8D-SE of SE

The Polytopic Roadmap describes the complexification process. It is the tool to explore and articulate future needs and trends for different development stages.

It outlines the road: Descriptive, Adaptive, Evolvable, Self-Evolvable and SE of SE.

These correspond to dimensions: 0D, 1D, 2D, 4D and 8D.

The 0D stage is related to the first level of modeling, that of examination and description of the objects or states denoted by K0. The next stage, 1D is that of learning or adaptive models. Learning or adaptability implies adjustment on the time scale of the existence of a system. It offers a preliminary and low dimensional perspective for complexity running (Iordache 1987).

The learning model describes processes in which the subject is repeatedly exposed to an experimental situation that is a condition or an event, for which various responses are possible, and each trial can modify the subject's response tendencies (Iosifescu and Grigorescu 1990).

The 1D frame from Fig. 1.2 illustrates in a condensed form the learning process represented as a segment from states K0 to conditions K1.

The next modeling stage shown in Fig. 1.2 stage 2D engages different levels of evolvability. The evolvability requires more than a process of learning or adaptation.

It needs innovation that is the capacity for change to autonomous advance into new life cycles, for instance new type of products, new market niches, new organizations and new conditioning levels. It requires process of processes. This entails a higher dimensional perspective and also the systems closure as shown in Fig. 1.2 stage 2D.

The evolvability refers to several levels, to a transformation from states or objects, K0, towards conditions, or processes K1, meta-conditions, or processes of processes K2, meta meta conditions, or processes of processes of processes K3 and finally to closed evolvable systems. Four levels systems in which the states K0 are related to a hierarchy of conditions K1, K2, K3 have been identified in different engineering domains were evolvability features have been emphasized (Iordache 2009, 2010, 2011).

Figure 1.2 stage 2D illustrates the basic cognitive frame for evolvability in which the states K0 interact in specific ways with the conditions K1, meta conditions K2, and meta meta conditions K3. Critical to achieve evolvability is the connection between the top levels of conditions K3 and the states K0. Artificially or naturally evolvable systems are supposed to cross this critical gap between the two extreme levels.

As shown in Fig. 1.2 following the Direct way, of convergence K0 $\rightarrow$ K1 $\rightarrow$ K2 $\rightarrow$ K3 we need to look at the Reverse way, of divergence K3$'$ $\rightarrow$ K2$'$ $\rightarrow$ K1$'$ $\rightarrow$ K0$'$.

The challenge is to put together the Direct and Reverse sequences. To do this we need 4D. We need to include a control capability between Direct and Reverse sequences.

We use the minuscule k0, k1 and so on to denote the control of Direct sequence K0, K1 and so on. We use the minuscule k0', k1' and so on to denote the control of Reverse sequence K0', K1' and so on.

The reference architecture shown in Fig. 1.2 stage 4D includes Direct and Reverse ways and for this reason is associated to conservative or sustainable systems. It describes the successive occurrence of convergence and divergence sequences. The Direct sequence of evolution is questioned and modified by the Reverse sequence.

The convergence way, represented in Fig. 1.2 stage 4D as a clockwise cycle corresponds to the integrating or relational approach to systems. The divergence way represented in Fig. 1.2 stage 4D as a counter-clockwise cycle corresponds to the reductionist or analytical approach to systems.

The ever growing complexity imposes paradigms as that of self-evolvability.

This focus on systems that self-configure, self-optimize, self-control, self-heal, self-direct, self-evolve and so on, based on a set of higher-level intrinsic capabilities and meeting of the user-specified variable objectives (Iordache 2012, 2013, 2017).

The polytope shown in Fig. 1.2 stage 4D, illustrates the reference architecture of self-evolvable systems. It is the 4D-cube and consists of a real cube and a control cube.

The control cube is a digital twin of the real cube (Glaessgen and Stargel 2012; Kritzinger et al. 2018; Hu et al. 2021). The 4D-cube is obtained by connecting every corner of the control cube, with the corresponding corner of the real cube. The tick hyper-line in 4D frame is a simplified representation of the connection of the two cubes (Thiel 1994). This kind of representation is useful for 8D polytope too.

The control capability may provide means for dialogue, and coordination of the systems.

Starting from 2012 the roads towards higher dimensional, beyond 4D have been emphasized (Iordache 2012, 2013).

The general term assigned to 4D and beyond 4D architectures was polytopic.

The polytopic projects will respond to the query to link reality to virtuality for materials, biosystems, cognitive systems, and for intelligent systems as engineering and mathematical ones (Iordache 2018, 2021).

The entire pattern of representation shown in Fig. 1.2 is significant for reference architecture evolution. A segment is made by two connected points, a square by two segments connected, a cube is made of two squares, front and back, with corners connected, and the 4D-cube, is made of two cubes, with corresponding corners connected.

The framework 8D is resulting as 4D of 4D architectures.

Grafting strategy that maintains the 0D vertices of the 4D architecture but replace any 0D vertices by developed and structured 4D vertices should be considered.

The intelligent-like behavior and sustainability by flourishing or grafting was studied by Ehrenfeld (2008) who defines it as the possibility that different life forms will flourish forever. Flourishing is understood as an emergent quality of the global socioeconomic and natural systems. (Ehrenfeld 2012; Ehrenfeld and Hoffman 2013).

By 4D hypercube flourishing to a hypercube of hypercubes that is an 8D framework is resulting from 4D framework. This is a strategy of grafting.

The new dimension is 8D = 4D + 4D. The first 4D come from the initial 4D frame as shown in Fig. 1.2. The next 4D come from the component 4D frames replacing the dots of the initial hypercube by the corresponding 4D frames. This strategy directs to the 8D Program.

1.4 Frameworks

The 4D frame is considered as the essential module for self-evolvable reference architectures. This challenges and replaces the 2D modules traditionally used as reference architectures. It is a change from squares to hypercubes, from 2 to 4D.

The doubling by iteration processes could be continued indefinitely but in any stage the duality feature should be maintained since this duality is the source for self-evolvability that is for live-like and then for intelligent-like capabilities.

Different presentations of the 4D and 8D will be analyzed in the following.

Figure 1.3 shows another presentation for the polytope 4D for complexity shown in Fig. 1.2.stage 4D (Iordache 2013, 2017).

The first and second cubes of 4D frame from Fig. 1.2 have been reunited as external and internal cubes.

The polytope 4D outlines that in interaction with the converging, constructing or Direct way K0 → K1 → K2 → K3 we need to look at the diverging, deconstructing or Reverse way K3′ → K2′ → K1′ → K0′.

The swinging from the Direct way to the dual Reverse way is critical since the boundaries where new information is created consist of coexisting tendencies of unification and diversification.

Since dual complementary descriptions are intrinsically irreducible and in some cases contradictory, the concepts integration or coordination of complementary descriptions requires a higher level description, a new dimension that emerges as a new hierarchical level of complexity. It is the 4D and beyond 4D reference architecture.

The central control cube, the Self, allows dialogue and interaction of the two ways in duality relation. These ways correspond to construction or Direct way, from K0 to K3, or to deconstruction or Reverse way, from K3′ to K0′.

The Self needs to endorse the dialogue, to ensure coordination and coherence of evolution on different levels. It results from Fig. 1.3 that the structure of the inner cube is isomorphic to that of the outer cube.

The inner cube is a digital twin of the outer cube (Rosen et al. 2015, Hartmann and Van der Auweraer 2021). Observe that the two cubes of 4D frame are reunited in a compact way. The 4D frame may be considered as two scaled.

Fig. 1.3 Polytope 4D for complexity-conservative strategy

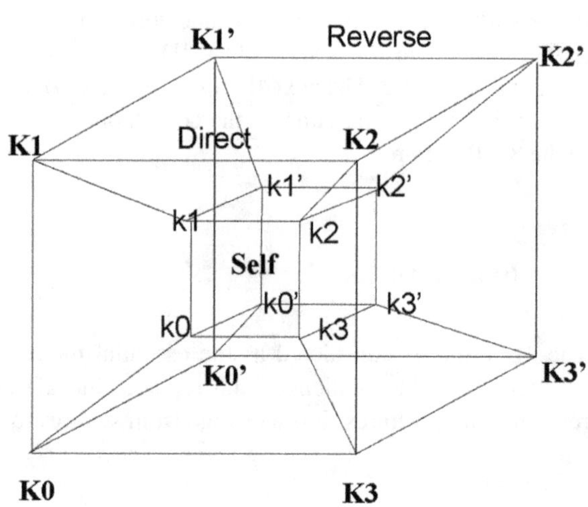

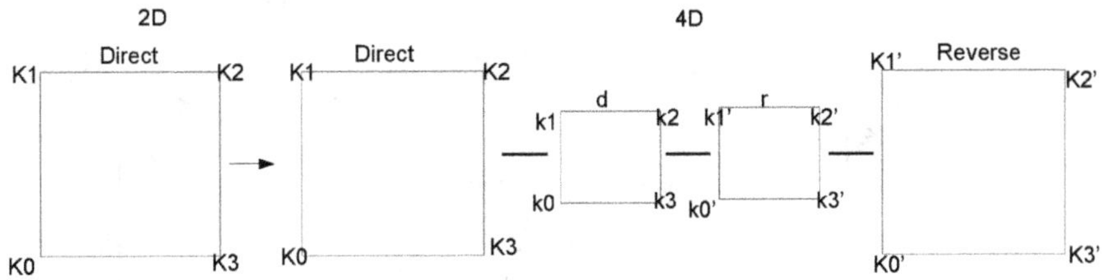

Fig. 1.4 Developed roadmap for polytope 4D for complexity

After any step in the Direct sequence, the systems make contact with the Self represented by the inner cube. Signals as k0, k1, k2 and k3 reinforce the evolution according to the Direct sequence. The set (k0, k1, k2, k3) corresponds to the control of Direct sequence.

If the Self have to change the direction shifting to k3', k2'.k1', k0' the next step will follow the Reverse sequence. The set (k0', k1', k2', k3') corresponds to the control of Reverse sequence.

Polytopes as presented in Fig. 1.3 have a general applicability as reference architectures since they take into account the inherent duality of complex systems.

Observe that duality is presented for both cubes the outer and the inner one.

Figure 1.4 shows a developed roadmap for the polytope 4D for complexity.

Only the 2D and 4D frames are presented here in a developed form.

The 2D of 2D step generates four squares as shown in Fig. 1.1 too.

Two correspond to the Direct and Reverse frames and two to the Self.

The 4D frames are based on Direct and Reverse sequences each considered as 2D frame. In this step the new frame, the Self is outlined.

The Self is shown with Direct d, part and Reverse r, part. Since Direct and Reverse frames are 2D, the d and r frames are 2D too. They illustrate the control of conditions.

Relating d and r allows obtaining the inner 3D-Cube.Finally the 4D frame, by inclusion the inner 3D frame in the external 3D frame. It will be a 4D-cube representing the self-evolvable systems.

Figure 1.5 shows more details of the polytope 8D for complexity.

This is the GRAF studied in detail as the objective of the 8D Program.

The grafting strategy that maintains the 4D architecture but replace any 0D vertices by developed and structured 4D vertices have been considered. It results a 4D hypercube of 4D hypercubes. The new dimension is 8D = 4D + 4D. The first 4D come from the initial 4D frames as shown in Fig. 1.3 and the next 4D from the component 4D frames replacing the dots of the initial hypercube. Hypercubes denoted by K0, K1 and so on may refer to real conditions or stages while hypercubes denoted by k0, k1 and so on may refer to virtual control.

Figure 1.5 results from Fig. 1.3 by grafting strategy.

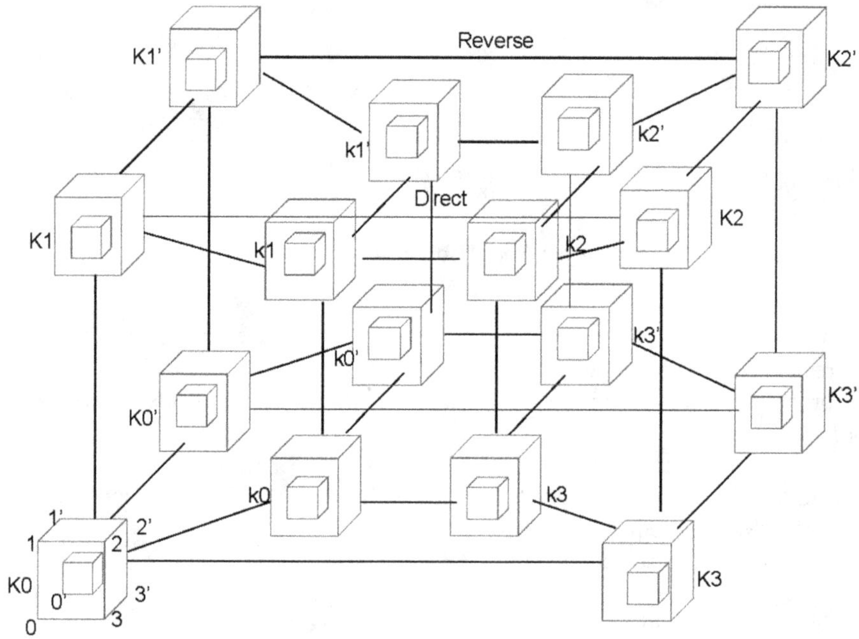

Fig. 1.5 Polytope 8D for complexity-conservative strategy

The sub-levels in K0 may be denoted by 0, 1, 2, 3 for a direct sub-sequence in K0, and by 3', 2', 1', 0' for a reverse sub-sequence in K0. For a complete description of K0, the sub-self of K0 components should be identified.

The presented Roadmaps outlined two sequences for evolution the Direct sequence: K0 → K1 → K2 → K3 that corresponds to the developmental stages and the Reverse sequence: K3′ → K2′ → K1′ → K0′ that corresponds to the dual modified states.

The apostrophe 'marks the dual or modified state.

This is a conservative or sustainable type of development. The directly acquired development is questioned and modified by the reverse sequence. The bottom-up construction from direct sequence complements the top-down deconstruction of the reverse sequence.

The conventional strategy is based on four stages K0, K1, K2 and K3.

In some cases we need to consider five stages, including the Self that centers the frame.

Figure 1.6 illustrates the developed roadmap for the polytope 8D for complexity.

It refers to the transition from 4 to 8D.

The 4D and 8D frames are presented here. The initial 4D architecture shows the developed frame of an external and an internal cube. These cubes result by relating the corresponding 2D frames.

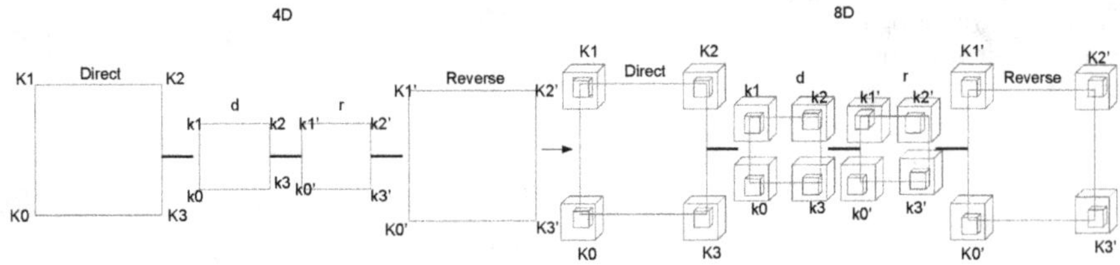

Fig. 1.6 Developed roadmap for polytope 8D for complexity

The transition from 4 to 8D is achieved by replacing any 0D vertex from the initial 4D architecture by component 4D frames. It is an iteration step that keeps the initial 4D architecture but replace any 0D vertices by component 4D vertices. We can do this only if we know what 4D vertices means.

It results a 4D hypercube of 4D hypercubes. The new dimension is 8D = 4D + 4D. The first 4D come from the initial 4D frame and the next 4D from the component 4D frames replacing the dots of the initial 4D frame.

In the next step an inclusion of a new frame, Self is considered. The Self is shown with Direct d, part and Reverse r, part.

Relating d and r allows obtaining the inner 7D-Cube. Finally the 8D frame results, by inclusion of the inner 7D-Cube in the outer 7D-Cube. The inner cube is a digital twin of the outer cube.

Figure 1.7 shows the Polytopic Roadmap for complexity with an innovative strategy.

Innovative strategy may replace conservative strategy.

In this case after the Direct sequence K0 → K1 → K2 → K3 we need to look at the Reverse way of divergence: K4 → K5 → K6 → K7.

The innovative type of roadmap that maintains the direct and the reverse senses to ensure the global equilibrium, but radically change the stages of reverse sequence should be considered as the candidate for future investigations.

The 4D frame in Fig. 1.7 is obtained by connecting every corner of the control cube, with the corresponding corner of the real cube. The tick line in 4D frame is a simplified representation of the connection of the two cubes.

Figure 1.8 shows another presentation and more details for Polytope 4D for complexity-innovative strategy.

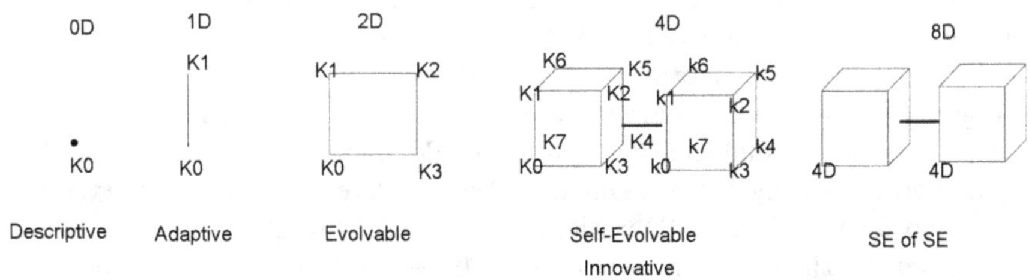

Fig. 1.7 Polytopic roadmap for complexity-innovative strategy

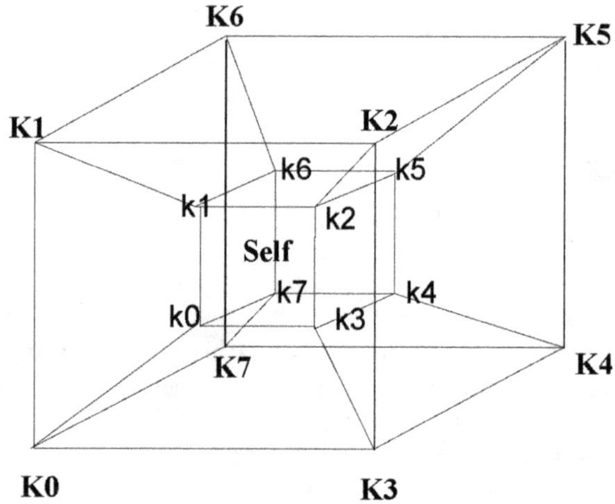

Fig. 1.8 Polytope 4D for complexity-innovative strategy

For innovative development we need eight different stages instead of four.

For innovative development the Direct sequence is: K0 → K1 → K2 → K3 while the Reverse sequence will be: K4 → K5 → K6 → K7.

The Reverse sequence is not just a deconstruction but a reconstruction. Reconstruction implies deconstruction but also going forward.

A continuous development involving all the stages would be: K0 → K1 → K2 → K3 → K4 → K5 → K6 → K7.

Finally K7 is in the same position relative to K0 as it was K0' in conservative Roadmaps.

Cycles are replaced by spirals.

The center or Self may be completed to the sequence: k0 → k1 → k2 → k3 → k4 → k5 → k6 → k7.

In some cases this center may became unnecessary. The transition from Direct sequences to Reverse sequences is possible without any central coordination. A procedure in which the center plays its role for conservative periods and became ineffective for innovative periods is of interest.

Arriving in K3 the system has to choose between closing the loop, establishing a center and modifying the achieved stages according to the sequence: $K3' \rightarrow K2' \rightarrow K1' \rightarrow K0'$ or search for an innovative sequence: K4 → K5 → K6 → K7.

This mechanism is valid for major stages. Moreover a similar strategic option conservative versus innovative may be considered for any sub-stage.

Figure 1.9 shows the innovative polytope 8D.

Figure 1.9 shows more details of the polytope 8D for complexity.

This is the GRAF studied in detail as the 8D Program.

The grafting strategy that maintains the 4D architecture but replace any 0D vertices by developed and structured 4D vertices is considered. It results a 4D hypercube of 4D hypercubes. The new dimension is 8D = 4D + 4D. The first 4D come from the initial 4D frame as shown in Fig. 1.9 and the next 4D from the component 4D frames

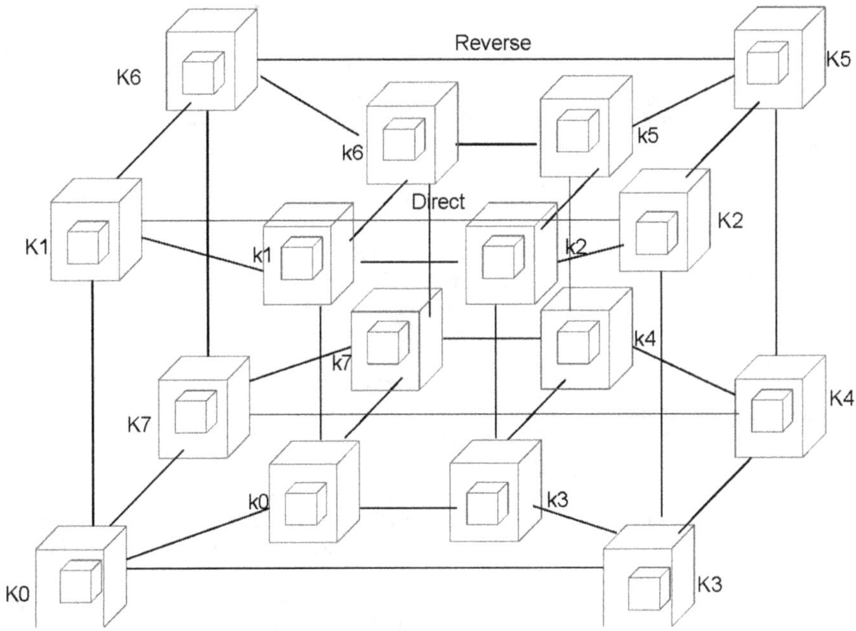

Fig. 1.9 Polytope 8D for complexity-innovative strategy

replacing the dots of the initial hypercube. Hypercubes denoted by K0, K1 and so on may refer to real conditions or stages while hypercubes denoted by k0, k1 and so on may refer to virtual control.

Figure 1.9 results from Fig. 1.8 by grafting strategy.

The presented Roadmaps outlined two sequences for evolution the Direct sequence K0 → K1 → K2 → K3 that corresponds to the developmental stages and the Reverse sequence K4 → K5 → K6 → K7 that corresponds to the innovative stages. The directly acquired development is questioned and innovated by the reverse sequence. The bottom-up construction from direct sequence complements the top-down innovation of the reverse sequence.

References

Adami, C.: What is complexity? BioEssays **24**(12), 1085–1094 (2002)

Cowan, N.: The magical number 4 in short-term memory: a reconsideration of mental storage capacity. Behavioral and Brain Sciences. **24**, 87–185 (2000)

Coxeter, H.S.M.: Regular Polytopes, 3rd edn. Dover, New York (1973)

Ebrahimi-Fard, K., Guo, L.: On products and duality of binary, quadratic, regular operads. J. Pure Appl. Algebr. **200**, 293–317 (2005)

Ehrenfeld, J.R.: Sustainability by Design: A Subversive Strategy for Transforming our Consumer Culture. Yale University Press, New Haven, CT (2008)

Ehrenfeld, J. R.: Beyond the brave new world: Business for sustainability. In: Bansal, P., Hoffman, A.J. (eds.) The Oxford Handbook of Business and the Natural Environment, pp. 611–619. Oxford: Oxford University Press (2012)

Ehrenfeld, J.R., Hoffman, A.J.: Flourishing: A Frank Conversation About Sustainability. Stanford University Press, Stanford, CA (2013)

Gardiner, J., Overall, R., Marc, J: The fractal nature of the brain: EEG data suggests that the brain functions as a "quantum computer" in 5–8 dimensions. Neuroquantology **8**, 137–41 (2010)

Glaessgen, E., Stargel, D.: The digital twin paradigm for future NASA and U.S. Air Force vehicles. In: Structures, Structural Dynamics, and Materials Conference, pp. 1–14 (2012)

Hartmann, D., Van der Auweraer, H.: Digital twins. In: Progress in Industrial Mathematics: Success Stories, pp. 3–17. Springer, Cham (2021)

Hu, W., Zhang, T., Deng, X., Liu, Z., Tan, J.: Digital twin: a state-of-the-art review of its enabling technologies, applications and challenges. Journal of Intelligent Manufacturing and Special Equipment **2**(1), 1–34 (2021)

Huang, H.-M., Messina, E.R., Wade, R.L., English, R.W., Novak, B., Albus, J.S.: Autonomy measures for robots. In: Proceedings of the International Mechanical Engineering Congress (IMECE), pp. 1–7 (2004)

Iordache, O.: Polystochastic Models in Chemical Engineering. VNU Science Press, Utrecht (1987)

Iordache, O.: Evolvable Designs of Experiments. Applications for Circuits, Wiley-VCH, Weinheim (2009)

Iordache, O.: Polystochastic Models for Complexity Springer, Berlin (2010)

Iordache, O.: Modeling Multi-Level Systems. Springer, Berlin (2011)

Iordache, O.: Self-Evolvable Systems. Machine Learning in Social Media, Springer, Berlin (2012)

Iordache, O.: Polytope Projects. Taylor & Francis, CRC Press, Boca Raton, USA (2013)

Iordache, O.: Implementing Polytope Projects for Smart Systems. Springer, Cham (2017)

Iordache, O.: Advanced Polytopic Projects. Springer, Cham (2018)

Iordache, O.: Polytopic Roadmaps. Springer, Cham (2021)

Iosifescu, M., Grigorescu, S.: Dependence with Complete Connections and Applications. Cambridge University Press, Cambridge (1990)

Kauffman, S.: At Home in the Universe: The Search for Laws of Self-Organization and Complexity. Oxford University Press, Oxford (1995)

Kritzinger, W., Traar, G., Henjes, J., Sihn, W., Karner, M.: Digital twin in manufacturing: a categorical literature review and classification. IFAC-Papers On Line **51**(11), 1016–1022 (2018)

Leroux, P.: On some remarkable operads constructed from Baxter operators (2003). arxiv:math. QA/0311214

Loday, J.-L., Vallette, B.: Algebraic Operads. Springer, Heidelberg (2012)

Mainzer, K.: Thinking in complexity, the complex dynamics of matter, mind and mankind. Springer, Berlin (1996)

Miller, G.A.: The magical number seven, plus or minus two: some limits on our capacity for processing information. Psychol. Rev. **63**(2), 81–89 (1956)

Muller, G.: A Reference Architecture Primer, White Paper. Eindhoven University of Technology, Eindhoven (2012)

Reimann, M.W., Nolte, M., Scolamiero, M., Turner, K., Perin, R., Chindemi, G., Dłotko, P., Levi, R., Hess, K., Markram, H.: Cliques of neurons bound into cavities provide a missing link between structure and function. Front. Comput. Neurosci. **11**, 48 (2017)

Reimann, M.W., Riihimäki, H., Smith, J.P., Lazovskis, J., Pokorny, C., Levi, R.: Topology of synaptic connectivity constrains neuronal stimulus representation, predicting two complementary coding strategies. PLoS ONE **17**(1) (2022)

Rosen, R., Von Wichert, G., Lo, G., Bettenhausen, K.D.: About the importance of autonomy and digital twins for the future of manufacturing. IFAC-Papers On Line **28**(3), 567–572 (2015)

Rossiter, N., Heather, M.: Four-level architecture for closure. In: Interoperability, EFIS2003, 5th International Workshop on Engineering Federated Information Systems, Coventry, UK, 17–18 July 2003, pp. 83–88 (2003)

Schafer, R.D.: An Introduction to Nonassociative Algebras. Academic Press Inc., New York (1966)

Springer, T.A., Veldkamp, F.D.: Exceptional groups. In: Octonions, Jordan Algebras and Exceptional Groups. Springer Monographs in Mathematics. Springer, Berlin (2000)

Thiel, T.: The design of the connection machine. Des. Issues **10**, 1, 5–18 (1994)

Chapter 2
Methods

Abstract A methodology to explore high complexity is introduced here. It is based on the Logic of Contradiction developed by Lupasco. This organizes the duality frames associated to the events E and non-events E'. Relation to the systems of thinking frames, System1 and System 2 as defined in Dual Process Theory, DPT, is highlighted. A list of Included Middle, IM, exemplars for different domains of reality is proposed. Their role for the dialogue in duality is outlined. Roadmaps, frameworks and roads to higher dimensionality for contradictions and complexity are presented. The Method to achieve higher dimensionality and explore complexity is described by the equation: Method = Iteration + Inclusion.

2.1 Principles

A logical system based on the dialectics of energy and applicable to high complexity real processes at different levels of reality was developed by Lupasco starting from 1935 (Lupasco 1947, 1951, 1960; Nicolescu 2002, 2010; Brenner 2010).

The Logic of Contradiction outlines the principle of antagonistic dualism and the principle of included third.

According to Lupasco every phenomenon or process, element or event E is always associated with a non-phenomenon, non-element or non-event E', such that the actualization of E entails the potentialization of E' and vice versa, alternatively, without either disappearing completely. The point of equilibrium- partial-actualization and partial-potentialization is one of maximum antagonism or contradiction from which an included third state, T emerges, resolving that contradiction.

If a process is possible in reality its counter process or non-process is also possible.

If by experiment we know only one side of the processes, the real one, we can construct by reason the other side of the process, the virtual one.

For non-living, material systems the process and the non-process may not exist in one place but somewhere in the universe. It is the case of matter and anti-matter.

For biological, cognitive and intelligent systems the process and the counter process may exist within. This is the root of control and autonomy in live-like and

O. Iordache, *General Reference Architecture Frameworks*, Studies in Systems, Decision and Control 73, https://doi.org/10.1007/978-3-031-70718-6_2

intelligent-like systems as considered in the study of Polytopic Roadmaps (Iordache 2021).

Duality and adjunction offer significant sources for contradiction frames, E, E'. The apostrophe 'marks the dual or modified state.

The duality for cognitive reference architecture was studied in dual process theory, DPT, (Wason and Evans 1975; Stanovich 1999, 2010; Kahneman 2003, 2011). The objective of DPT was to understand how the cognitive processes develop, how the dynamical organization of the brain structure can be viewed in a computational context and to propose multi-purposes polytopic reference architectures.

The cognitive polytopic reference architectures show multiple levels and sub-levels. There exists multi-level reference architectures that may be characterized by dual ways of divergence and convergence processes organized in a space for dialogue.

Dual processing accounts of cognition have been emphasized in a range of areas including adaptive learning, reasoning, designing, decision making, clinical diagnostic and social cognition (Kahneman 2003, Evans 2008, Croskerry 2009, McGilchrist 2009).

Processes that are rapid, automatic and effortless on the one hand are contrasted with those that are slow, sequential and controlled on the other. We characterize these two types of processes as independent sources of control for behavior that may come into conflict and contradiction.

These dual processes have been mapped on two distinct cognitive systems. These two systems have been given various characterizations including emotional-rational, heuristic-analytic, implicit-explicit, associative and rule-based and more general System 1, S1 and System 2, S2 (Stanovich and West 2000; Kahneman 2003).

System 1, S1, using this as label for the fast, automatic or non-standard is often described as evolutionarily old, and independent of individual differences of intelligence levels, whereas System 2, S2, is by contrast slow, standard, evolutionarily recent, and related to heritable differences in intelligence and working memory capacity.

System 1 and System 2 may be associated to convergent and divergent systems respectively in the cognitive reference architectures. Despite contradiction and competition the dialogue and cooperation of the dual ways S1 and S2 is mandatory.

Studying complementarity and contradiction, some authors draws on the biology of the human brain, namely, the dual nature of its hemispheric specializations. The left and right hemispheres have relatively distinct cognitive functions, and the reality, as perceived and communicated by the human brain, is a complementary union of opposites (Ji 1995; Springer and Deutsch 1998; McGilchrist 2009).

Table 2.1 contains some pairs of systems pertaining to contradiction.

Duality studies reveal forward and backward, composing and decomposing, construction and deconstruction, homogenization and heterogenization and other pairs of contradictions or antinomies.

The Included Middle, IM, labeled also as Self or Included Third, T, define the attempts to solve contradiction by dimension increasing.

Table 2.1 Contradiction

Event E	Non-event E'	Application domains
Assimilation	Accommodation	Development
Automatic	Controlled	Control
Concrete	Abstract	
Construction	Deconstruction	
Contextualized	Decontextualized	
Convergence	Divergence	Design
Cyclic	Linear	
Direct	Reverse	Logic of contradiction
Eastern thought	Western thought	
Engineering	Science	Engineering science
Extent	Intent	Formal concept analysis
Fast	Slow	
Innovation	Tradition	
Old	Recent	Evolution
Parallel	Serial	
Physical	Cyber	Cyber physical systems
Product	Coproduct	Hopf algebras
Real	Digital	Digital twin
Right hemisphere	Left hemisphere	
Society	Economy	Social market economy
Strategy	Tactics	
Synthetic	Analytic	
System 1, S1	System 2, S2	Dual process theory

Swinging mediated by the Included Middle between two complementary or contradictory ways is crucial for reuniting and complementing different trends, for self-evolvability.

For material level of reality catalysts and templates may play the role of the Included Middle. Similar roles are played by ribosome, enzymes, and cellular nucleus for biological level of reality.

Table 2.2 shows possible illustrations of the concept of Included Middle, IM, for different domains of reality including cognitive and intelligence domains.

They may be a source of inspiration for industrial, economical or social high complexity projects (Iordache 2022).

The Grothendieck topos concept arose in algebraic geometry, as a consequence of combining the concept of sheaf and closure under categorical operations. The subsequent developments associated with logic are more interdisciplinary and trans-disciplinary. The potential use of toposes as unifying bridges in mathematics and other domains has been emphasized by Caramello (2017).

Table 2.2 Included middle

Authors	Concept
Baars (2002)	Global workspace
Brier (2008, 2009)	Semantic mind
Caramello (2017)	Grothendieck topos
Ceulemans et al. (2003)	Core array
Cook (1986)	Corpus callosum
Dehaene et al. (2003)	Global neuronal workspace
Deleuze (1988, 1993)	Fold. rhizome
Ehreshmann (2012)	Archetypal core
Glaessgen and Stargel (2012)	Digital twin
Hillis (1985)	Connection machine
Houdé (2019)	Inhibitory control
Iordache (2012, 2013, 2017)	Self
Kelso and Egstrom (2006)	Metastable coordination
Kraft and Eibeck (2020)	Knowledge graph
Latour (1993)	Middle Kingdom
Lupasco (1947, 1951, 1960)	Included middle
Nicolescu (2002, 2010, 2015)	Hidden third
Overton (2013)	Relational metatheory
Pattee (1995, 2001)	Semantic closure
Peirce (1956)	Abduction
Piaget (1970, 1973)	Post formal
Rizzolatti and Craighero (2004)	Mirror neurons
Sporns (2013)	Rich club
Tononi and Edelman (1998)	Dynamic core
von Uexkull (1973)	Inner world
Wilber (2000)	Vision logic
Worrall (1989)	Structural realism

An early definition for the Digital Twin, as proposed by Glaessgen and Stargel, (2012), describes it as a multi-physics, multi-scale, and probabilistic simulation of a complex product that uses models, sensor updates, and so on, to mirror the life of its corresponding physical twin.

The Connection Machine, is a type of computing engine which computes through the interaction of many, simple identical processing/memory cells. Because the processing takes place concurrently, the machine can be much faster than a conventional computer (Hillis 1985).

A knowledge graph is a network of data expressed as a directed graph, where the nodes of the graph are concepts or their instances and the edges of the graph are links between related concepts or instances. Knowledge graphs provide a powerful means to host, query and traverse data, and to find and retrieve related information (Kraft and Eibeck 2020; Akroyd et al. 2021).

Lupasco (1951) and Houdé (2019) go beyond the approaches of Kahneman emphasizing (System 1, S1) and Piaget emphasizing (System 2, S2) to put forward a theory in which the included third T, or the inhibitory-control system (System 3, S3) takes precedence.

T or S3 denotes the interface or the bridge between S1 and S2. Houdé argues that the brain contains a third control system located in the prefrontal cortex which is dedicated to inhibiting intuitive heuristics, non-standard system, S1, and activating Piaget's logical algorithmic, standard system, S2, anywhere in the brain on a case-by-case basis, depending on the goal and context of the task. The "triple-S" system S1, S2, S3 with possible identifications S1 Society, S2 Science, S3 Self was considered by Sellberg et al. (2021).

Concepts as Included Middle, IM, (Lupasco 1951, 1960) and Hidden Third, HT, (Nicolescu 2002, 2010, 2015) proved to be significant for high complexity studies.

After postulating evolvability Lupasco is going further and refers to self-evolvable that is live-like systems. A being is composed when a process and a counter process, that is a non-process form a relational loop. To achieve self-evolvability, the middle should be included in the larger sustainable frame. The resulting non-Aristotelian logic is the Logic of the Included Middle, consisting of axioms and rules of inference for determining the state of the dynamic elements involved in a being phenomenon.

The reflecting Included Middle is for Lupasco the site of the absolute contradiction.

It is pure contradiction between heterogeneity and homogeneity. Its products, the concepts are themselves ruled by the Logic of the Contradiction. They are dynamic elements in which an extensive value dominated by the homogeneous stays in contradiction with the intensive value characterized by its heterogeneous essence.

The concept, as opposed to what the classical logic claims, is the locus of the contradiction (Lupasco 1960).

Following Lupasco, Nicolescu takes a transdisciplinary approach to nature and knowledge. Transdisciplinary knowledge, as based on the so-called Logic of Hidden Third, necessarily leads to the inclusion of values and systemic understanding versus simple analytic reasoning. According to this logic, in every relation involving two complementary levels of experience, there is a third level that belongs simultaneously to both. Since complementary descriptions, as that for E and E', from Lupasco theory may appear as antagonistic or contradictory the concept and physical integration or coordination of the descriptions requires a higher level description, a new dimension that emerges as a new hierarchical level of complexity, as a center and a higher dimension in the same time (Nicolescu 2010). The more complex is a system the more dimensions and polytopic reference architectures are necessary to manage high complexity (Iordache 2013, 2018, 2021). Dimension increases by iteration and inclusion, to take into account the contradiction of contradictions inherent to the Logic of Contradiction. So, we are confronted with a non-Aristotelian, perspective (Korzybski 2023).

Peirce (1956) suggested that there is a form of inference, abduction, besides deduction and induction, and that abduction is the only type of inference capable of producing new ideas. His work is related to processes of scientific discovery, to conjectures, where abduction was a stage of forming hypotheses to explain surprising facts.

Mirror neurons are a particular class of visuomotor neurons that discharge both when the primate does a particular action and when observes another individual doing a similar action (Rizzolatti and Craighero 2004). Mirror neurons represent the neural basis of a mechanism that creates a direct link between the sender of a message and its receiver between virtual and real (Goertzel et al. 2008). For this reason they may be considered as support of Included Middle.

Worrall (1989) consider that structural realism steers a middle path between scientific realism and antirealism.

2.2 Roadmaps

Figure 2.1 shows the Polytopic Roadmap, PTR, for contradiction.

Notations are: Events-E and Non-Events, E'.

E0, E1, E2 and E3 denotes the possible Events sequence, while E0', E1', E2' and E3' denotes the possible non-Events sequence.

We use the minuscule e0, e1, e2, e3 to denote the control of Direct sequence E0, E1, E2, E3. We use the minuscule e0', e1', e2', e3' to denote the control of Reverse sequence E0', E1', E2', E3'.

The dimensions illustrated in Fig. 2.1 are:

- 0D-States
- 1D-Processes
- 2D-Evolvable
- 4D-Self-Evolvable
- 8D-SE of SE

States are considered as 0D. Processes result from States as 1D segment. The Process includes States in duality, for instance initial and final States. Several Processes build an Evolvable systems are represented here as 2D.

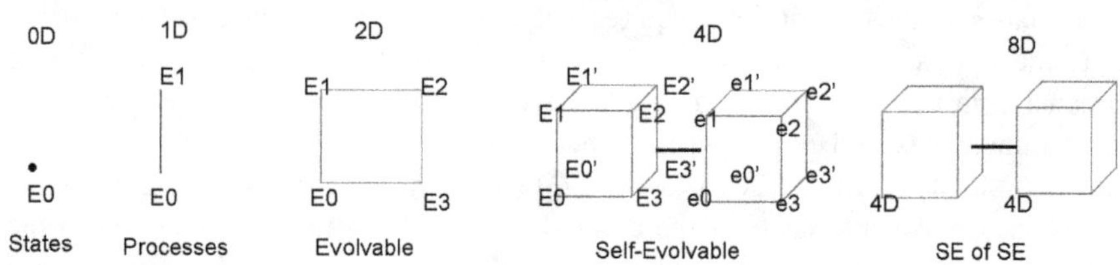

Fig. 2.1 Polytopic roadmap for contradiction

The Self-Evolvable, 4D, includes Processes in duality. The Event E imposes the non-Event E'.

The 4D polytope shown in Fig. 2.1 illustrates the architecture of potentially self-evolvable systems. It is a representation consisting from a pair of 3D-cubes. The second cube is a digital twin of the first one.

The 4D-cube is obtained by joining all corners of the first 3D-cube with the corresponding corners of the second 3D-cube. The 8 lines connecting the two 3D-cubes are represented in Fig. 2.1 by a thick hyper-line. In this way the 4D-cube has a simplified representation. This kind of representation is useful for 8D polytope too (Thiel 1994).

The thick hyper-line connects the two 7D-cubes generating the 8D polytope.

The second 7D-cube is a digital twin of the first 7D-cube.

The Polytopic Roadmap shown in Fig. 2.1 is based on the Lupasco principle of antagonistic dualism and the principle of included third. It outlines the principle of antagonistic dualism in the transition from 0 to 1D, to 2D and the principle of included third in the steps from 2 to 4D and from 4 to 8D.

The 8D is resulting as 4D of 4D architecture that is as SE of SE.

Grafting or flourishing strategies that maintain the 0D vertices of the 4D architecture but replace any 0D by developed and structured 4D frames are considered. They are partially represented in Fig. 2.1 to avoid drawing complication. This kind of grafting was studied by Ehrenfeld (2008) who defines sustainability as the possibility that different life forms will flourish forever. Flourishing is an emergent quality of the global socioeconomic and natural systems (Ehrenfeld and Hoffman 2013).

By flourishing or grafting the 4D hypercube it results a hypercube of hypercubes that is an 8D framework. The new dimension is 8D = 4D + 4D. The first 4D in the sum come from the initial 4D frame as shown in Fig. 2.1 and the next 4D from the component 4D frames replacing the dots of the initial hypercube. This grafting allows achieving 8D that is, an intelligent-like level.

The Polytopic Roadmap from Fig. 2.1 illustrates the road construction after the 2D stage. The resulting 4D frame is considered as the essential module for self-evolvable reference architectures. The doubling by iteration processes could be continued indefinitely but in any stage the duality feature should be maintained since this duality is the root for self-evolvability that is live-like and for intelligent-like levels.

2.3 Frameworks

Different presentations of the 4D and 8D will be analyzed in the following.

The first and second cubes of 4D frame shown in Fig. 2.1 may be reunited as external and internal cubes.

Figure 2.2 shows details of the polytope 4D for contradiction. The Logic of Contradiction is presented as Events E (Direct sequence) and Non-Events E' (Reverse sequence), the forth and the back face of the outer cube.

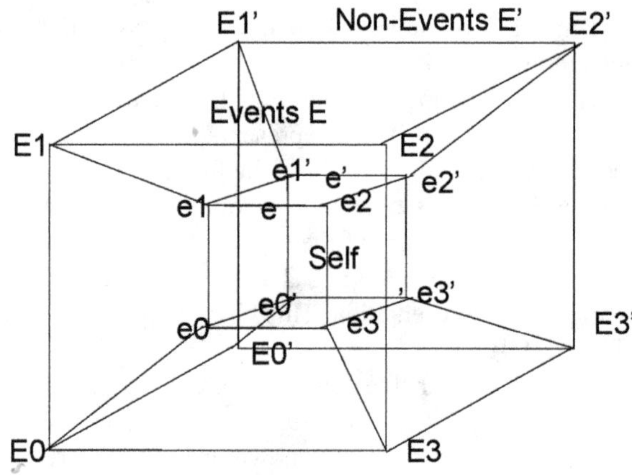

Fig. 2.2 Polytope 4D for contradiction

E0, E1, E2 and E3 denote the Events corresponding to different levels of conditioning.

E0', E1', E2' and E3' denote the Non-Events corresponding to different levels of reconditioning.

Here e and e' marks the activation of the corresponding faces of the inner cube. This cube corresponds to Self or in other words to Included Middle, IM, or to Included Third, T.

The inner cube is a digital twin of the outer cube.

After any step in the Direct sequence, the system contact the Self represented by the inner cube. Signals as e0, e1, e2 and e3 reinforce the evolution according the Events sequence. The set (e0, e1, e2, e3) correspond to the control of direct sequence. If the Self have to change the direction shifting to e3', e2', e1', e0' the next step will be along the reverse Non-Event sequence. The set (e0', e1', e2', e3') corresponds to the control of Reverse sequence.

Observe that duality is presented for both cubes the outer and the inner one.

The inner cube is a digital twin of the outer cube.

Figure 2.3 shows a developed roadmap for polytope 4D for contradiction.

It is a developed version of some parts of Polytopic Roadmap, PTR.

The 2D and 4D frames are presented in a developed form here.

The 2D frames are based on Events E sequences. E impose E'.

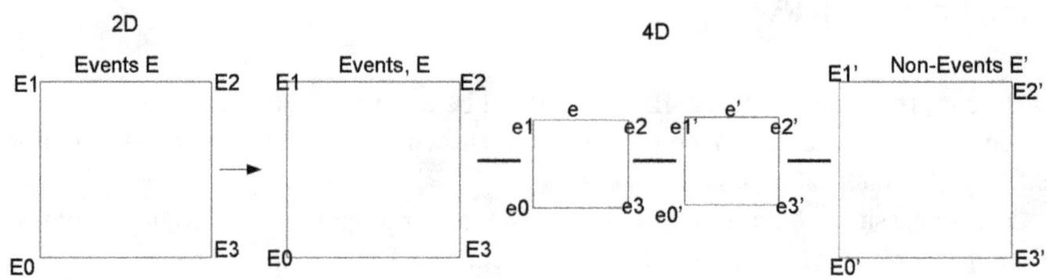

Fig. 2.3 Developed roadmap for polytope 4D for contradiction

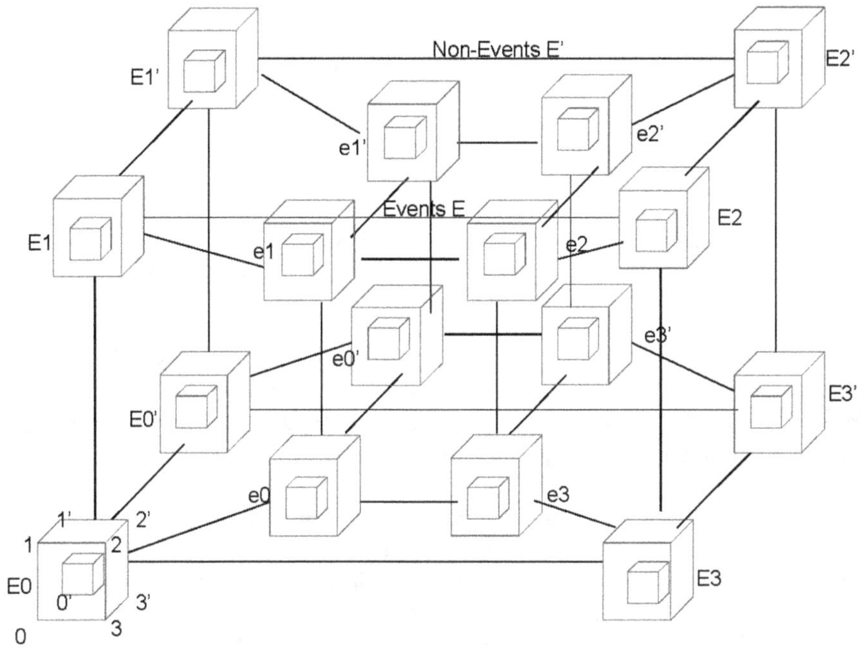

Fig. 2.4 Polytope 8D for contradiction

In the next step the new frame, the Self or Included Middle should be considered.

The Self is shown with Events e, part and Non-Events e', part. Since E and E' frames are 2D the e and e' frames should be 2D too. They illustrate the conditions.

Relating e and e' allows obtaining the inner cube. Finally the 4D frame results by inclusion the inner cube frame in the external cube frame. It will be a 4D-cube representing the Self-Evolvable (SE) systems.

Figure 2.4 shows the Polytope 8D for contradiction.

It results from Fig. 2.2 by grafting. This replaces the 0D from Fig. 2.2 by 4D frames.

The sub-levels in E0 may be denoted by 0,1,2,3 for a direct sub-sequence in E0, and by 3', 2', 1', 0' a reverse sub-sequence in E0. For a complete description of E0, the sub-self of E0 components should be identified.

The presented Roadmaps outlined two sequences of events for evolution the direct sequence: E0 → E1 → E2 → E3 that corresponds to the Events and the reverse sequence: E3′ → E2′ → E1′ → E0′ that corresponds to the Non-Events. This is a sustainable type of development. The directly acquired development is questioned and modified by the reverse sequence. The bottom-up construction from direct sequence complements the top-down deconstruction of the reverse sequence.

Hypercubes denoted by E0, E1 and so on may refer to real events while hypercubes denoted by e0, e1 and so on refer to virtual control.

Figure 2.5 shows a developed roadmap for polytope 8D for contradiction.

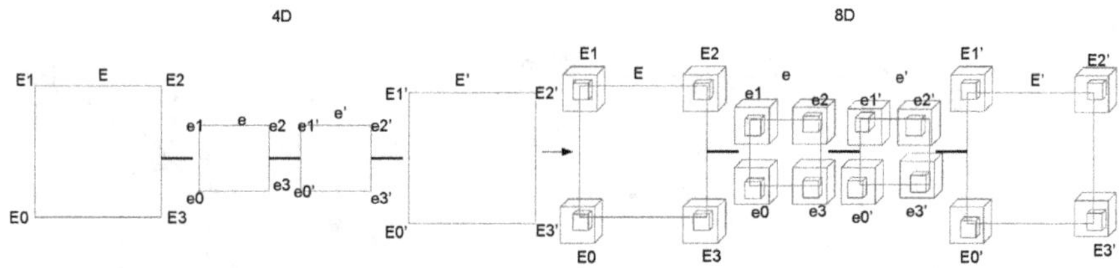

Fig. 2.5 Developed roadmap for polytope 8D for contradiction

The 4D and 8D frames are presented in Fig. 2.5.

The initial 4D architecture shows the developed frame of an external and of an internal cube. The transition from 4 to 8D is achieved by replacing any 0D vertex from the initial 4D architecture by component 4D frames. It is the grafting strategy that keeps the initial 4D architecture but replace any 0D vertices by component 4D vertices.

It results a 4D hypercube of 4D hypercubes. The new dimension is 8D = 4D + 4D. In this sum, the first 4D come from the initial 4D frame and the next 4D from the component 4D frames replacing the dots of the initial 4D hypercube.

An inclusion of a new frame called the Included Middle or Self is considered.

The Self is shown with Events e, part and Non-Events e', part. Hypercubes denoted by E0, E1 and so on refer to real events while hypercubes denoted by e0, e1 and so on refer to virtual control.

Relating e and e' allows obtaining the inner 7D-Cube. The 7D-Cube is a 3D-Cube with hypercubes 4D as vertices. Finally the 8D frame results, by inclusion of the inner 7D-cube in the external 7D-Cube. The inner cube is a digital twin of the outer cube.

2.4 Method

Higher dimensional frames may be shaped starting from dots, 0D, using these to generate the 1D segments, using these to generate 2D squares, and so on Roadmaps have been proposed as Methods to organize knowledge to understand and exceed complexity.

They outline Aristotelian and non-Aristotelian features (Korzybski 2023).

Aristotle developed a logic providing us with concepts such as the law of identity and the excluded middle, EM.

In his Discourse on Method, Descartes explored the basic laws of thinking, and fashioned them into the foundations for inquiry. Descartes elaborate a Method and rules for the direction of the mind providing us with an orientation for the way we think a focus on reduction, simplification, and clarity. What Descartes proposed as rules for the direction of mind have become the foundation for considered good thinking, and institutionalized. There we find the same, the increasing specialization in departments, the splitting up into smallest possible parts, and the creation of strong boundaries based on the three axioms of classical Aristotelian logic.

Several attempts to describe and formalize knowledge reference architectures served as inspiration for Polytopic Roadmaps.

Grothendieck (1983) in pursuing stacks studied the sequence: objects, 1-morphisms, 2-morphisms, 3-morphisms, and so on (Lurie 2009). Baas (2015, 2016) in studying hyperstructures refers to the sequence: 0-bounds, 1-bounds, 2-bounds, 3 bonds and so on.

The Object Management Group (OMG), a computer industry standards consortium refers to the sequence: data, models, meta models, meta meta models and so on (Mellor 2004, OMG 2008).

These examples are general theories having objective similar to that of polystochastic models and polytopic projects (Iordache 2010, 2013, 2017, 2018, 2021).

For polystochastic models the studied roadmap was: states, processes, processes of processes, processes of processes of processes or in other words, states, processes, meta processes, meta meta processes and so on.

The highlight is on processes rather than on states or objects.

The research direction of polystochastic models and of polytopic projects is based on the idea of construction by iteration of processes of processes over processes and so on. The processes of different levels will show specific properties. This means that for different stages we have to consider iterations in new conditions and not just recursive procedures.

Table 2.3 shows more examples of iterations methods (Winiwarter 1986a, b; Iordache 2021).

Abbreviations are: EDoE-Evolvable Designs of Experiments (Iordache 2009), GST-General Systems Theory (von Bertalanffy 1968), HC-hypercycles (Eigen 1971), HS-hyperstructures (Baas 2009), IRS-Iosifescu Random Systems (Iosifescu and Grigorescu 1990), MES-Memory Evolutive Systems (Ehresmann 2012), MOF-Metal–Organic Frameworks (Yaghi 2019), PSM—Polystochastic Models (Iordache 2010), RIFS-Random Iterated Function Systems (Barnsley 1993), SM-Supramolecules (Lehn 1995), SoS-Systems of Systems (Maier 1998).

Table 2.3 backs the research for Methods to organize knowledge to understand and to explore high complexity domains. The Methods are based on iteration process followed by an attempt to solve contradiction by increasing dimension by inclusion.

In abbreviated form we have the equation: Method = Iteration + Inclusion.

This equation is founded on the Lupasco principle of antagonistic dualism and the principle of included third (Lupasco 1951).

It is valid for different levels of reality. Hartmann (1952) considered four reality levels: material, biologic, cognitive and intelligent. For these levels Morin (1977, 1985, 2003, 2014) described his Method as: nature of nature, life of life, knowledge of knowledge and humanity of humanity.

An operad is a structure that consists of abstract operations, each one having a fixed finite number of inputs (arguments) and one output, as well as a specification of how to compose these operations. The operad describes operations of operations as a family

Table 2.3 Iteration method

Definition of iteration	Perspective	References
Categories of categories	n-categories, MES	Lurie (2009), Ehresmann (2012)
Control of control	Cybernetics	Wiener (2019)
Cybernetics of cybernetics	2nd order cybernetics	von Foerster (2003)
Cycles of cycles	Hypercycles, HC	Eigen (1971)
Designs of designs	EDoE	Iordache (2009)
Discipline of disciplines	Transdisciplinarity	Jantsch (1970), Piaget et al. (1972)
Distinction of distinctions	Forms	Spencer-Brown (1979)
Evolution of evolution	Self-organization	Nicolis and Prigogine (1977)
Forms of forms	Catastrophes	Thom (1972)
Functions of functions	Fractals, RIFS, IRS	Mandelbrot (1982), Barnsley (1993)
Hierarchy of hierarchies	Holons	Koestler (1967)
Humanity of humanity	Complexity	Morin (2003)
Knowledge of knowledge	Complexity	Morin (2014)
Life of life	Complexity	Morin (1985)
Models of models	Meta models	Mellor (2004), OMG (2008)
Nature of nature	Complexity	Morin (1977)
Networks of networks	Hypernetworks	Johnson (2014)
Operation of operations	Operads	May (1972), Leinster (2004)
Order of orders	Social order	von Hayek (1975)
Organization of organization	Synergetics	Haken (2000)
Processes of processes	Polystochastic models	Iordache (2010)
Production of production	Autopoiesis	Maturana and Varella (1980)
Proofs of proofs	Metasystems	Nagel and Newman (1958)
Reality of reality	Communication	Watzlawick (1976)
Society of societies	Communication	Luhmann (1997), Minsky (1988)
States of states	IRS, RIFS	Iosifescu and Grigorescu (1990)
Structures of structures	Hyperstructures, HS	Baas (2009, 2015, 2016)
SM of SMs, MOF of MOFs	Polytopic chemistry	Lehn (1995), Yaghi (2019)
Systems of systems	GST, SoS	von Bertalanffy (1968), Maier (1998)

of composable operations satisfying several intuitive properties like associativity of composition and permutability of the inputs (May 1972; Leinster 2004).

The society of societies focuses on communication as the only genuinely social phenomenon (Luhmann 1997). Social systems differentiate and evolve as they communicate in three separate dimensions: the social, temporal, and functional. Communication, bit by bit, produces social structures that, recursively, produce future structures. Society is communication. Sociology, as the science of society, is communication about how different societal systems operate, communicate, evolve, and maintain their boundaries. Luhmann's systems theory has been applied in different scientific disciplines, such as sociology, political science, jurisprudence, sociology of law, therapy, educational science, literature, and communication research. There are many connections to similar theoretical approaches, like second-order cybernetics, chaos theory, radical constructivism, and other systems theories.

Minsky (1988) proposes that the human mind is composed of a society of agencies, each of which is composed of a society of smaller agencies, and so on. Mind is a society of societies. While the lower-level agencies could not be considered as living or intelligent, their composition into structured wholes yields behavior that is markedly live-like and intelligent-like. Such self-similar or recursive structures emerge naturally and are efficient for processing information (Spivak 2013).

Principle of antagonistic dualism and iteration supports the presence of events E and non-events E'. It is a dimension increase from that of E to that of the frame (E E'). For instance if E and E' frames are 2D then (E E') frame is 3D. It should be emphasized that it is an Iteration accompanied by duality and maintenance of the antagonistic dualism.

The 3D is unsustainable in some cases.

According to the equation: Method = Iteration + Inclusion, to the Iteration, we add the Inclusion that is the Included Middle step, T, increasing again in this way the dimension (Lupasco 1951, 1952; Nicolescu 2002, 2010; Brenner 2008, 2010).

The new higher dimensional frame can be denoted (E T E'). It is a 4D frame.

The Method highlights the roads towards high dimensional reference architectures.

Table 2.4 shows a synthesis of Method applications in complexity studies.

Different situations are characterized as Standard or non-Standard.

The term **non-standard** typically refers to something that deviates from the usual or accepted norm. It can apply to various contexts, such as models, processes or practices.

The challenge was to put together Systems and non-Systems. The label poly was associated to the label multiple encountered in high complexity studies.

Multiple closures refer to Cartesian splits as: mind–body, object-environment, internal external, object-subject and so on (Primas 1994; Atmanspacher 1997; Pattee 2001).

Category theory consistently seen by much of the mathematical community as "abstract nonsense" or "ridiculously abstract" is today itself under attack for relatively few applications outside mathematics.

Labeling an argument as "abstract nonsense", is not intended to be derogatory. Instead it may be considered as a caveat for the generality and originality. The use of categorical concepts proves to be a natural part of transferring and integrating the knowledge across disciplines. The restructuring employed in applied category theory helps to elucidate common themes across disciplines. The drive for a common language and comparison of similar structures in algebra and topology is what led to the development of category theory in the first place, and recent hints show that this approach is not only useful between mathematical disciplines, but between sciences and engineering or more general between virtuality and reality. Further applications

Table 2.4 Synthesis of method applications

Systems (S2) standard	Non-systems (S1) non-standard	Complexity roots	ReferencesIordache	Domain
Markovian	Non-Markovian	Multiple processes Multiple levels	(1987, 2010)	Polystochastic
Archimedean	Non-Archimedean	Multiple fields Multiple scales	(2011)	Complexity
Cartesian	Non-Cartesian	Multiple dimensions Multiple closures	(2012, 2013)	pOlytopic Higher dimension
Aristotelian	Non-Aristotelian	Multiple logics Multiple disciplines	(2022)	Higher dimension Transdisciplinarity
"Sense" sets	"Non-Sense" categories	Multiple theories and technologies		Diversify unify Virtuality-reality

are hampered by the widespread view of category theory as exceedingly and unnecessarily abstract. However, it should be emphasized that the generality is necessary as it allows for new connections to formal methods in sciences and engineering (Breiner et al. 2019).

Figure 2.6 shows the roads to higher dimensional contradiction.

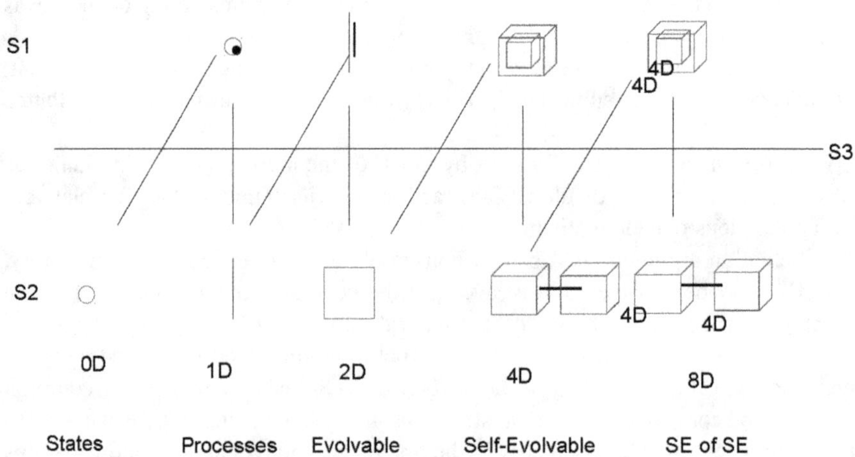

Fig. 2.6 Roads to higher dimensional contradiction

S2 and S1 roads have been outlined.

S2 is based on Aristotelian logic while S1 support a non-Aristotelian logic (Korzybski 2023).

Consider that the S2 road starts from a State, 0D, represented here as a circle. Included in this an S1 frame appears. It is associated to upward S3 traverse and is represented as an included dot. The dot and the circle will be related in a segment or line, 1D, as shown on the S2 road. The contradiction is present if one of extremities of the segment is a source or an entrance while another is a sink or an exit. The segment corresponds to Processes. These are oriented.

After a new upward transition of the S3, on the S1 road an included middle represented as a segment appears. Transition to S2 road imposes the development of both segments. By relating the two segments the 2D square results.

The contradiction is still present if one segment that is one side of the square is oriented upside and the opposed side is oriented downside. The 2D square is associated to Evolvable systems.

On the S1 road an included cube appears and develops. It results the 4D hypercube on the S1 road. Transition to S2 road imposes the development and the correlation of two cubes. The 4D-cube of Self-Evolvable systems results. The cubes maintain contradiction. The Direct and Reverse tracking directions for front and back faces for one cube is coupled to complementary Reverse and Direct tracking for the other cube. This ensures Self-Evolvability. The 4D-cube on the S2 road is obtained by joining all corners of the first 3D-cube with the corresponding corners of the second 3D-cube. The 8 lines connecting the two 3D-cubes are represented in Fig. 2.6 by a thick hyper-line. In this way the 4D-cube has a simplified representation.

This representation is useful for 8D polytope too.

The dimensions shown in Fig. 2.6 are:

- 0D-State
- 1D-Processes
- 2D-Evolvable
- 4D-Self-Evolvable
- 8D-SE of SE

The 8D framework is resulting as 4D of 4D architecture.

This corresponds to Self-Evolvability of Self-Evolvability, SE of SE.

The process may continue indefinitely in the same way as a dance or a zigzag between S1 and S2 roads. The transitions through S3 increase the dimension.

An examination of the S1 road show included frames: dot included in a dot, segment included in a segment, square included in a square, and so on. They are included and this corresponds to the non-Aristotelian S1, traits.

On the S2 road it is the sequence of outer frames. On the S2 rows we observe: two dots defining a segment, two parallel edges of a square, two cubes of a hypercube and so on.

On this road we don't have inclusions but exclusions and this corresponds to the Aristotelian logic.

These S1 and S2 roads may be linked to recapitulation theories. As example we refer to the familiar hypothesis in biology, attributed to Haeckel, that ontogeny (S1) recapitulates phylogeny (S2). It has also been outlined by studies of the emergence of life that recapitulates the appearance of matter (Ulanowicz 2009).

Piaget and Garcia scientific methodology starts from the hypothesis that there exists a parallelism between a particular problem solving and the historical development of the involved sciences (Piaget and Garcia 1989). The short history of an individual problem solving, that is the problem ontogeny, is considered as parallel to the evolutionary long history of a lineage that is the problem phylogeny.

That reality portrayed as the balance between complementary or contradictory trends has long been an element of traditional Eastern philosophy (Mair 1990; Xu et al. 2018).

This view has been promoted also in Western science. A general approach is the Logic of Contradiction or Logic in Reality (Lupasco 1952; Brenner 2008).

According to this vision the roads to higher dimensions appear as a zigzag between the centrifugal expansiveness of entropy on the S2 road and the centripetality of self-reinforcing information on the S1 road. It is an iteration process and an attempt to solve the contradictions by dimension increasing and inclusion (Xu et al. 2018).

References

Akroyd, J., Mosbach, S., Bhave, A., Kraft, M.: Universal digital twin – A dynamic knowledge graph. Data-Centric Eng. 2 (2021)

Atmanspacher, H.: Cartesian cut, Heisenberg cut, and the concept of complexity. World Futur.: J. Gen. Evol. **49**(3–4), 333–355 (1997)

Baars, B.J.: The conscious access hypothesis: origins and recent evidence. Trends Cogn. Sci. **6**, 47–52 (2002)

Baas, N.A.: Hyperstructures, topology and datasets. Axiomathes **19**, 281–295 (2009)

Baas, N.A.: Higher order architecture of collections of objects. Int. J. Gen. Syst. **44**(1), 55–75 (2015)

Baas, N.A.: On higher structures. Int. J. Gen. Syst. **45**(6), 747–762 (2016)

Breiner, S., Sriram, R., Subrahmanian, E.: Compositional models for complex systems. In: Lawless, W., Mittu, R., Sofge, D., Moskowitz, I.S., Russell, S. (eds.) Artificial Intelligence for Internet of Everything, pp. 241–270. Academic (2019)

Brenner, J.E.: Logic in Reality. Springer Science & Business Media: Berlin (2008)

Brenner, J.E.: The philosophical logic of Stéphane Lupasco (1900–1988). Log. Log. Philos. **19**, 243–284 (2010)

Brier, S.: Cybersemiotics: Why Information is Not Enough. University of Toronto Press, Toronto, Canada (2008)

Brier, S.: Cybersemiotic pragmaticism and constructivism. Construct. Found. **5**(1), 19–39 (2009)

Caramello, O.: Theories, Sites, Toposes: Relating and Studying Mathematical Theories Through Topos-Theoretic 'Bridges', Oxford University Press (2017)

Ceulemans, E., Van Mechelen, I., Leenen, I.: Tucker3 hierarchical classes analysis. Psychometrika **68**, 413–433 (2003)

Cook, N.D.: The Brain Code: Mechanisms of Information Transfer and the Role of the Center. Methuen, New York (1986)

Croskerry, P.: Clinical cognition and diagnostic error: applications of a dual process model of reasoning. Adv. Health Sci. Educ. **14**(1), 27–35 (2009)

Dehaene, S., Sergent, C., Changeux, J.P.: A neuronal network model linking subjective reports and objective physiological data during conscious perception. Proc. Natl. Acad. Sci. u.s.a. **100**, 8520–8525 (2003)

Deleuze, G.: Foucault. University of Minnesota Press, Minneapolis (1988)

Deleuze, G.: The Fold: Leibniz and the Baroque. University of Minnesota Press, Minneapolis (1993)

Ehrenfeld, J.R.: Sustainability by Design: A Subversive Strategy for Transforming our Consumer Culture. Yale University Press, New Haven (2008)

Ehrenfeld, J.R., Hoffman, A.J.: Flourishing: A Frank Conversation About Sustainability. Stanford University Press, Stanford, CA (2013)

Ehresmann, A.C.: MENS, an info-computational model for (neuro-) cognitive systems capable of creativity. Entropy **14**(9), 1703–1716 (2012)

Eigen, M.: Self-organization of matter and the evolution of biological macromolecules. Naturwissenschaften **58**, 465–523 (1971)

Evans, J.St.B.T.: Dual-processing accounts of reasoning, judgment and social cognition. Annu. Rev. Psychol. **59**, 255–278 (2008)

Glaessgen, E., Stargel, D.: The digital twin paradigm for future NASA and U.S. Air Force vehicles. In: Structures, Structural Dynamics, and Materials Conference, pp. 1–14 (2012)

Goertzel, B., Aam, O., Smith, F.T., Palmer, K.: Mirror neurons, mirrorhouses, and the algebraic structure of the self. Cybern. & Hum. Knowing **15**(1), 9–28 (2008)

Grothendieck, A.: Pursuing stacks (1983)

Haken, H.: Information and Self-Organization : A Macroscopic Approach to Complex Systems. Springer Series in Synergetics, 2nd edn. Springer, Berlin (2000)

Hartmann, N.: The New Ways of Ontology. Greenwood Press, Westport (1952)

Hillis, W.D: The Connection Machine. MIT Press (1985)

Houdé, O.: 3-System Theory of the Cognitive Brain, A Post-Piagetian Approach to Cognitive Development. Routledge, London (2019)

Iordache, O.: Polystochastic Models in Chemical Engineering. VNU Science Press, Utrecht (1987)

Iordache, O.: Evolvable Designs of Experiments. Applications for Circuits. Wiley-VCH, Weinheim, Germany (2009)

Iordache, O.: Polystochastic Models for Complexity. Springer, Berlin (2010)

Iordache, O.: Modeling Multi-Level Systems. Springer, Berlin (2011)

Iordache, O.: Self-Evolvable Systems. Machine Learning in Social Media. Springer, Berlin (2012)

Iordache, O.: Polytope Projects. Taylor & Francis, CRC Press, Boca Raton (2013)

Iordache, O.: Implementing Polytope Projects for Smart Systems. Springer, Cham (2017)

Iordache, O.: Advanced Polytopic Projects. Springer, Cham (2018)

Iordache, O.: Polytopic Roadmaps. Springer, Cham (2021)

Iordache, O.: Roads to Higher Dimensional Polytopic Projects-Reference Architectures. Springer, Cham (2022)

Iosifescu, M., Grigorescu, S.: Dependence with Complete Connections and Applications. Cambridge University Press, Cambridge (1990)

Jantsch, E.: Inter- and transdisciplinary university: a systems approach to education and innovation. Policy. Sci. **1**, 403–428 (1970)

Ji, S.: Complementarism: a biology-based philosophical framework to integrate western science and eastern Tao. In: Proceeding of the 16th International Congress of Psychoterapy, pp. 518–548 (1995)

Johnson, J.: Hypernetworks in the Science of Complex Systems. Imperial College Press, London (2014)

Kahneman, D.: A perspective on judgement and choice. Am. Psychol. **58**(9), 697–720 (2003)

Kahneman, D.: Thinking, Fast and Slow. Farrar, Straus and Giroux, New York (2011)

Kelso, J.A.S., Engstrom, D.A.: The Complementary Nature. MIT Press, Cambridge (2006)

Koestler, A.: The Ghost in the Machine. Hutchinson, London (1967)

Korzybski, A.: Science and Sanity: An Introduction to Non-Aristotelian Systems and General Semantics, Sixth edn. (2023)

Kraft, M., Eibeck, A.: J-Park simulator: knowledge graph for industry 4.0. Chem. Ing. Technol. **92**, 967–977 (2020)
Latour, B.: We Have Never Been Modern. Harvard University Press, Cambridge, MA (1993)
Lehn, J.-M.: Supramolecular Chemistry: Concepts and Perspectives. Wiley-VCH, Weinheim (1995)
Leinster, T.: Higher Operads, Higher Categories. Cambridge University Press, Cambridge (2004)
Luhmann, N.: Die Gesellschaft der Gesellschaft. Frankfurt A.M.: Surhkamp (1997)
Lupasco, S.: Logique et contradiction. Presses Universitaires de France, Paris (1947)
Lupasco, S.: Le principe d'antagonisme et la logique de l'énergie. Prolégomènes à une science de la contradiction. Hermann & Co. (1951)
Lupasco, S.: Le principe d'antagonisme et la logique de l'énergie. Rev. Métaphysique Et De Morale **57**, 467–468 (1952)
Lupasco, S.: Les trois matières. Julliard, Paris (1960)
Lurie, J.: On the classification of topological field theories. In: Current Developments in Mathematics, 2008, pp.129–280. International Press, Sommerville (2009)
Maier, M.W.: Architecting principles for system of systems. Syst. Eng.: J. Int. Counc. Syst. Eng. **1**(4), 267–284 (1998)
Mair, V. (ed.): Tao Te Ching: The Classic Book of Integrity and the Way. Bantam Books, NY (1990)
Mandelbrot, B.: The Fractal Geometry of Nature. W.H. Freeman, San Francisco (1982)
Maturana, H., Varela, E.: Autopoiesis and Cognition. The Realization of the Living. Reidel, Dordrecht (1980)
May, P.: The Geometry of Iterated Loop Spaces, vol. 271. Lecture Notes in Mathematichs. Springer, Berlin (1972)
McGilchrist, I.: The Master and His Emissary: The Divided Brain and the Making of the Western World. Yale University Press, New Haven (2009)
Mellor, S.J.: MDA Distilled: Principles of Model-Driven Architecture. Addison-Wesley Professional (2004)
Minsky, M.: The Society of Mind. Simon and Schuster (1988)
Morin, E.: La Methode 1: La Nature de la Nature. Seuil, Paris (1977)
Morin, E.: La Methode 2: La Vie de la Vie. Seuil, Paris (1985)
Morin, E.: La Methode 5: L'Humanité de l'humanité. L,identité humaine. Seuil, Paris (2003)
Morin, E.: La Methode 3: La Connaissance de la Connaissance. Pocket Book (2014)
Nagel, E., Newman, J.R.: Gödel's Proof. New York University Press, New York (1958)
Nicolescu, B.: Manifesto of Transdisciplinarity. SUNY, Albany, USA (2002)
Nicolescu, B.: Methodology of transdisciplinarity: levels of reality, logic of the included middle and complexity. Transdiscip. J. Eng. Sci. **1**, 19–38 (2010)
Nicolescu, B.: The hidden third as the unifier of natural and spiritual information. Cybern. Hum. Knowing **22**(4), 91–99 (2015)
Nicolis, G., Prigogine, I.: Self-Organization in Nonequilibrium Systems. Wiley, New York (1977)
OMG: Object Management Group, Software & Systems Process Engineering Meta Model Specification 2.0 (2008)
Overton, W.F.: Relationism and relational developmental systems: A paradigm for developmental science in the post-Cartesian era. Adv. Child Dev. Behav. **44**, 21–64 (2013)
Pattee, H.H.: Evolving self-reference: matter, symbols, and semantic closure. Commun. Cogn.–Artif. Intell. **12**(1–2), 9–28 (1995)
Pattee, H.H.: the physics of symbols: bridging the epistemic cut. Biosystems **60**, 5–21 (Special Issue: Reflections on the work of Howard Pattee) (2001)
Peirce, C.S.: Collected Papers of Charles Sanders Peirce, vols. 1–8. Cambridge University Press, Cambridge (1956)
Piaget, J.: Genetic Epistemology. Columbia University Press, New York (1970)
Piaget, J.: The epistemology of interdisciplinary relationships. In: Briggs, A., et al. (eds.) Interdisciplinarity, Problems of Teaching and Research in Universities, pp. 127–139. OECD, Paris (1972)

Piaget, J.: Introduction a l'epistemologie genetique. La pensee Mathematique. Presses Universitaires de France, Paris France (1973)
Piaget, J., Garcia, R.: Psychogenesis and the History of Science. Columbia University Press, New York (1989)
Primas, H.: Endo- and exotheories of matter. In: Atmanspacher, H., Dalenoort, G.J. (eds.) Inside Versus Outside. Springer, Berlin (1994)
Rizzolatti, G., Craighero, L.: The mirror-neuron system. Annu. Rev. Neurosci. **27**, 169–192 (2004)
Sellberg, M.M., Cockburn, J., Holden, P.B., Lam D.P.M: Towards a caring transdisciplinary research practice: Navigating science, society and self. Ecosyst. People **17**, 292–305 (2021)
Spencer-Brown, G.: Laws of Form. Allen and Unwin, London (1979)
Spivak, D.I.: The operad of wiring diagrams: formalizing a graphical language for databases, recursion, and plug-and-play circuits (2013). arxiv.org/abs/1305.0297
Sporns, O.: Network attributes for segregation and integration in the human brain. Curr. Opin. Neurobiol. **23**(2), 162–171 (2013)
Springer, S.P., Deutsch, G.: Left Brain, Right Brain, 5th edn. W. H. Freeman, New York (1998)
Stanovich, K.E.: Who is Rational? Studies of Individual Differences in Reasoning. Mahway, NJ, Lawrence Elrbaum (1999)
Stanovich, K.E.: Rationality and the Reflective Mind: Oxford University Press (2010).
Stanovich, K.E., West, R.F.: Individual difference in reasoning: implications for the rationality debate. Behavioral and Brain Sciences **23**(5), 645–726 (2000)
Thiel, T.: The design of the connection machine. Des. Issues **10**, 1, 5–18 (1994)
Thom, R.: Structural Stability and Morphogenesis. W. A. Benjamin, Reading, Mass (1972)
Tononi, G., Edelman, G.M.: Consciousness and complexity. Science **282**, 1846–1851 (1998)
Ulanowicz, R.E.: A Third Window: Natural Life beyond Newton and Darwin. Templeton Foundation Press, West Conshohocken (2009)
von Foerster, H.: Cybernetics of cybernetics. In: von Foerster H. (eds.) Understanding Understanding, pp. 283–286. Springer, New York (2003)
von Hayek, F.: Kinds of Order in Society, Studies in Social Theory, vol. 5. Institute for Humane Studies, Menlo Park (1975)
von Uexküll, J.: Theoretische Biologie. Frankfurt A. M.: Suhrkamp Taschenbuch Wissenschaft (1973)
Wason, P.C., Evans, J.St.B.T.: Dual processes in reasoning? Cognition **3**, 141–154 (1975)
Watzlawick, P.: How Real is Real? Communication, Disinformation, Confusion. Random House, New York (1976)
Wiener, N.: Cybernetics or Control and Communication in the Animal and the Machine. MIT Press (2019)
Wilber, K.: Integral Psychology. Shambhala Publication. Inc. (2000)
Winiwarter, P.: Autognosis: the theory of hierarchical self-image building systems. In: Proceedings of International Conference on Mental Images, Values & Reality. Society For General Systems Research (1986a)
Winiwarter, P.: Concepts of self-organization, self-organization of concepts. In: Proceedings of International Conference on Mental Images, Values & Reality. Society for General Systems Research (1986b)
Xu, Z., Cheng, G., Ulanowicz, R.E., Song, X., Deng, X., Zhong, F.: The common developmental road: tensions among centripetal and centrifugal dynamics. Nat. Sci. Rev. **5**, 417–426 (2018)
Yaghi, O.M.: Reticular chemistry in all dimensions. ACS Cent. Sci. **5**(8), 1295–1300 (2019)
Worrall, J.: Structural realism: the best of both worlds? Dialectica **43**, 99–124 (1989)

Chapter 3
Operations and Equipments

Abstract Combinatorial species are viewed as basic unit operations. Permutations, distributors collectors, middle vessel columns and separations are the case studies considered here. The similarities of the associated Polytopic Roadmaps are emphasized. Examples of involved mathematical tools are the dual graded graphs and the formal concept analysis. The 4D, that is live-like, and 8D, that is intelligent-like, frameworks have been presented in detail.

3.1 Permutations

Simple designs of experiments (DoE) matrices may be associated to different permutations.

Figure 3.1 shows equivalent notations for permutations

We associate the permutation 231 to the DoE matrix in which for any experiment just one factor is modified. Experiments are assigned to rows and factors to columns in that matrix.

For the study of hierarchy of experiments and factors we can use the permutation trees (Fomin 1992, 1994).

Figure 3.2 shows the dual graded graphs for permutation trees.

Figure 3.2a shows the U graph while Fig. 3.2b shows the D graph.

The interconnections between elements make the difference between U and D. The unique minimal element is denoted here by $\hat{0}\hat{0}$.

The permutation graphs may be considered as trees of elementary designs of experiments. Figure 3.3 shows the dual permutation trees of designs of experiments.

Figure 3.3a shows the U graph while Fig. 3.3b shows the D graph.

U corresponds to constructions or combinations, D to deconstructions or decompositions.

The procedure developed by Fomin can be considered as a basic process of doubling and relating. These are the procedures generating Polytopic Roadmaps.

U0, U1, U2 and so on denote different levels of the U-graph.

D0, D1, D2 and so on denote different levels of the D-graph.

O. Iordache, *General Reference Architecture Frameworks*, Studies in Systems, Decision and Control 73, https://doi.org/10.1007/978-3-031-70718-6_3

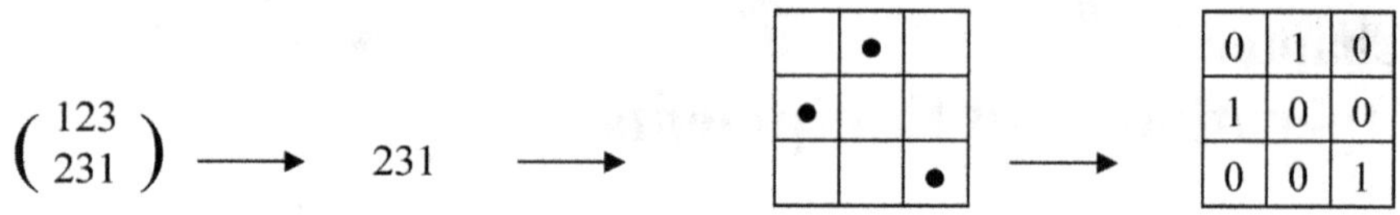

Fig. 3.1 Equivalent notations for permutations

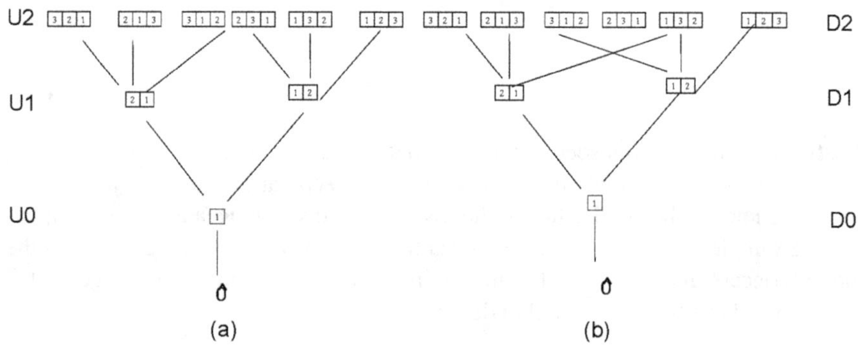

Fig. 3.2 Dual permutation trees

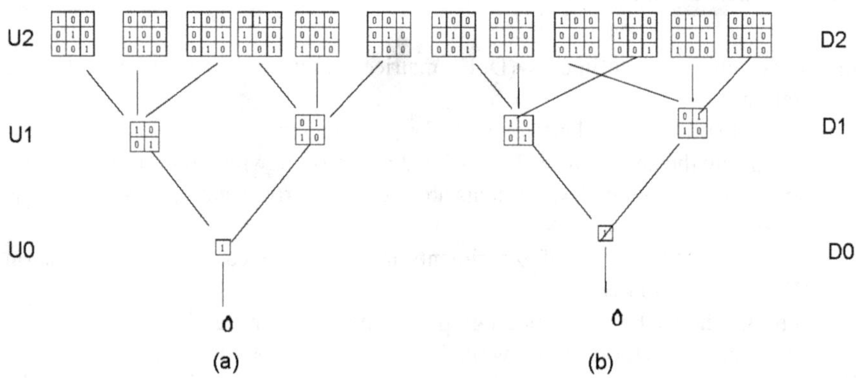

Fig. 3.3 Dual permutation trees of designs

The method to increase dimensionality based on dual graded graphs is illustrated. Figure 3.4 shows the Polytopic Roadmap, for permutations.

We use the minuscule u0, u1, u2, u3 to denote the control of Direct sequence U0, U1, U2, U3 of the U-graphs. We use the minuscule d0, d1, d2, d3 to denote the control of Reverse sequence D0, D1, D2, D3 of the D-graphs.

The dimensions illustrated in Fig. 3.4 are:

- 0D-States
- 1D-Processes
- 2D-Evolvable

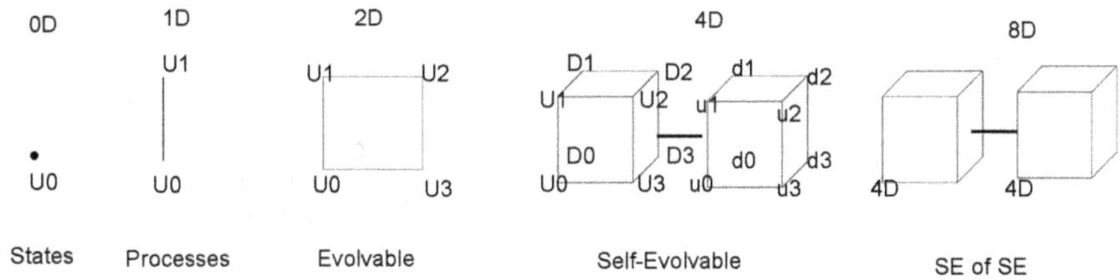

Fig. 3.4 Polytopic roadmap for permutations

- 4D-Self-Evolvable
- 8D-SE of SE

States are considered as 0D. Processes result from States as 1D segment. The Process includes States in duality, for instance initial and final States.

Several Processes build an Evolvable systems are represented here as 2D. The Evolvable includes Processes in duality.

The 4D polytope shown in Fig. 3.4 illustrates the architecture of potentially self-evolvable systems. It is a representation consisting from a pair of 3D-cubes.

The second cube is a digital twin of the first cube (Kritzinger et al. 2018).

The 4D-cube is obtained by joining all corners of the first 3D-cube with the corresponding corners of the second 3D-cube. It is the doubling followed by relating procedure. The 8 lines connecting the two 3D-cubes are represented in Fig. 3.4 by a thick hyper-line. In this way the 4D-cube has a simplified representation (Thiel 1994; Ammon 1998).

This representation is useful for 8D polytope too. A thick hyper-line connects the two 7D-cubes generating the 8D polytope. The second 7D-cube is a digital twin of the first 7D-cube.

Grafting strategies that maintains the 0D of the 4D architecture but replace any 0D vertices by developed and structured 4D vertices are considered. They are partially represented. This kind of grafting was discussed by Ehrenfeld (2008) who consider sustainability as the continuous possibility of life forms (Ehrenfeld and Hoffman 2013).

By 4D hypercube grafting it results a hypercube of hypercubes that is an 8D framework. The new dimension is 8D = 4D + 4D. The first 4D come from the initial 4D frame as that shown in Fig. 3.4 and the next 4D from the component 4D frames replacing the dots of the initial hypercube. This grafting allows achieving an intelligent-like level.

Figure 3.5 shows more details and a different presentation of the polytope 4D for permutations.

Here u and d marks the activation of the corresponding faces of the inner cube.

The inner cube corresponds to Self or in other words to Included Middle.

The inner cube is a digital twin of the outer cube.

After any step in the U-graph, the systems contact the Self represented by the inner cube. Signals as u0, u1, u2 and u3 reinforce the evolution according the U-graph.

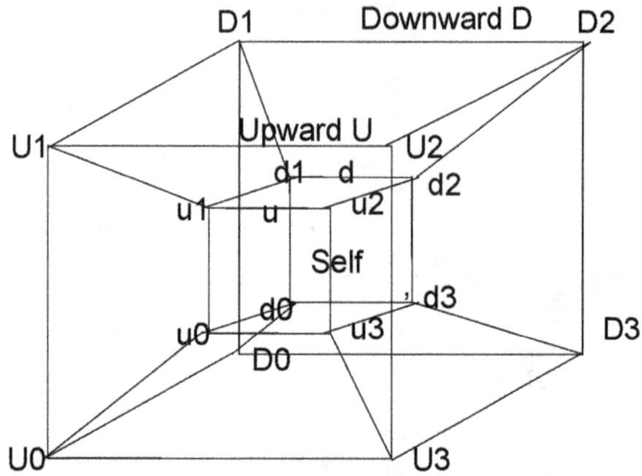

Fig. 3.5 Polytope 4D for permutations

The set (u0, u1, u2, u3) may correspond to the control of U-graph. If the Self have to change the direction shifting to d3, d2, d1, d0 the next step will be along the D-graph. The set (d0, d1, d2, d3) correspond to the control of D-graph.

Polytopes as presented in Fig. 3.5 have a general applicability as reference architectures since they take into account the inherent duality of complex systems.

Observe that duality is preserved for both cubes the outer and the inner one in Fig. 3.5.

Figure 3.6 shows the Polytope 8D for permutation.

It results from Fig. 3.5 by grafting.

The presented Roadmaps outlined two sequences of levels for evolution: U0 → U1 → U2 → U3 and: D3 → D2 → D1 → D0.

The levels may have a sub-level structure. The sub-levels are denoted 0, 1, 2, 3 in the direct sequence and 3', 2', 1', and 0' in reverse direction. Control frames should be considered at any level. This ensures that the control actions are distributed.

The 8D shows central and distributed control capabilities.

3.2 Distributors Collectors

Figure 3.7 shows distributor collector systems.

Fluid enters the distributor through a single inlet flow through the branching channels and leaves the distributor through a square array of outlets. This distributor could uniformly lead air over the catalytic layer of a fuel cell as well as distribute liquids. Such structures could feed high-throughput setups or uniformly heat or cool surfaces. They may be integrated into micro-fluidic devices (Coppens 2012).

The connection between distributor and collector is completed by the fluid surrounding the devices.

Observe that the level increasing is based on doubling.

C0, C1, C2 and C3 shows the levels for connectors.

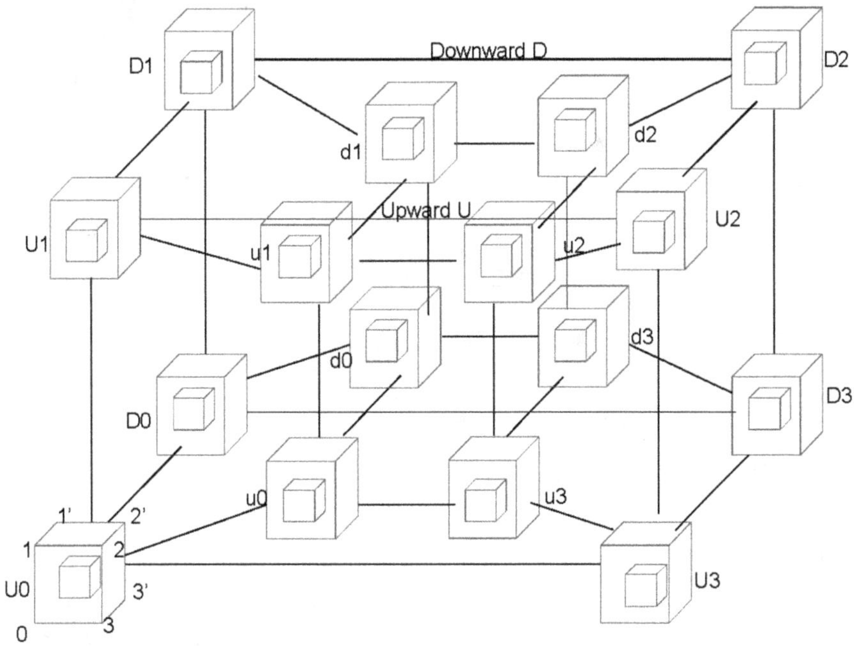

Fig. 3.6 Polytope 8D for permutation

D0, D1, D2 and D3 shows the levels for distributors.

Figure 3.8 shows the Polytopic Roadmap for distributors collectors.

Notations are C for collectors and D for distributors.

The dimensions illustrated in Fig. 3.8 are:

- 0D-States
- 1D-Processes
- 2D-Evolvable
- 4D-Self-Evolvable
- 8D-SE of SE

States are considered as 0D. Process results from States as 1D segment. The Process includes States in duality, for instance initial and final States. Several Processes build an Evolvable systems are represented here as 2D. The Evolvable frame includes Processes in duality. The Self controls the transition between D-distributors and C-collectors, both being 2D. It is a 4D frame.

We use the minuscule d0, d1, d2, d3 to denote the control of Direct sequence D0, D1, D2, D3. We use the minuscule c0, c1, c2, c3 to denote the control of Reverse sequence C0, C1, C2, C3.

The 4D polytope shown in Fig. 3.8 illustrates the architecture of potentially self-evolvable systems. It is a representation consisting from a pair of 3D-cubes.

The second cube is a digital twin of the first cube.

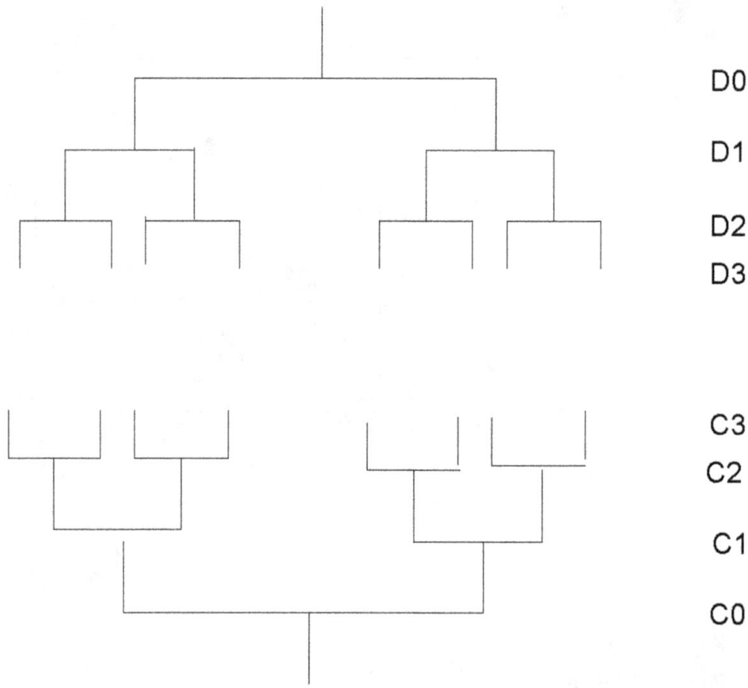

Fig. 3.7 Distributors collectors

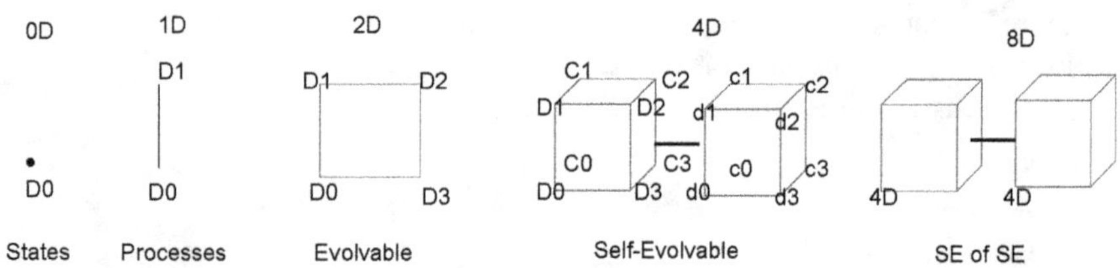

Fig. 3.8 Polytopic roadmap for distributors collectors

The 4D-cube is obtained by joining all corners of the first 3D-cube with the corresponding corners of the second 3D-cube. The 8 lines connecting the two 3D-cubes are represented in Fig. 3.8 by a thick hyper-line. In this way the 4D-cube has a simplified representation. This representation is useful for 8D polytope too. A thick hyper-line connects the two 7D-cubes generating the 8D polytope.

The second 7D-cube is a digital twin of the first 7D-cube.

The framework 8D is resulting as 4D of 4D architecture.

Grafting strategies that maintains the 0D of the 4D architecture but replace any 0D vertices by developed and structured 4D vertices are considered. In Fig. 3.8 they are partially represented to avoid drawing complication.

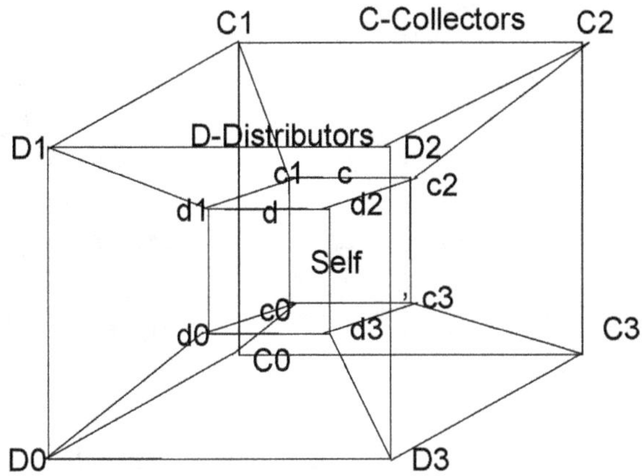

Fig. 3.9 Polytope 4D for distributors collectors

By 4D hypercube grafting it results a hypercube of hypercubes that is an 8D framework. The new dimension is 8D = 4D + 4D. The first 4D in the sum come from the initial 4D frame and the next 4D from the component 4D frames replacing the dots of the initial hypercube. This grafting allows attaining the 8D, intelligent-like level.

Figure 3.9 shows more details and another presentation of the polytope 4D for distributors collectors.

Here d and c marks the activation of the corresponding faces of the inner cube. This cube corresponds to Self or in other words to Included Middle.

After any step in the distributors D-graph, the systems make contact with the Self represented by the inner cube. Signals as d0, d1, d2 and d3 reinforce the evolution according the D-graph. The set (d0, d1, d2, d3) correspond to the control of D-graph.

If the Self have to change the direction by shifting to c3, c2, c1, c0 the next step will be along the C-graph. The set (c0, c1, c2, c3) correspond to the control of C-graph.

Figure 3.10 shows the Polytope 8D for permutations.

It results from Fig. 3.9 by grafting.

The presented Roadmaps outlined two sequences of levels D0 → D1 → D2 → D3 that corresponds to the D levels and C3 → C2 → C1 → C0 that corresponds to the C levels.

The levels may have a sub-level structure. The sub-levels are denoted 0, 1, 2, 3 in the direct sequence and 3', 2', 1', and 0' in reverse direction. Control frames have been considered at any level. In this way the control activities are distributed.

The 8D shows both, the central as the distributed control capabilities.

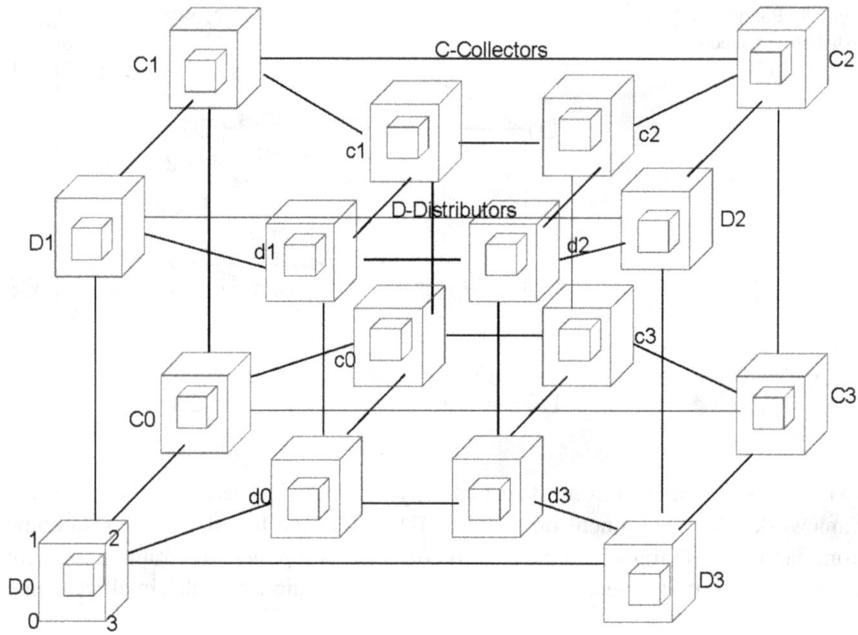

Fig. 3.10 Polytope 8D for distributors collectors

3.3 Middle Vessel Columns

Batch distillation is a commonly used separation process in the pharmaceutical and fine-chemical industries. Figure 3.11 illustrates batch distillation configurations.

The most frequently used batch distillation configuration is the batch rectifier (BR).

Figure 3.11a shows a batch rectifier (BR) configuration.

In the case of a batch stripper (BS) the vessel ensuring the feed arriving at the top stage of the column is located above the column (Fig. 3.11b).

Middle vessel column configuration (Fig. 3.12) received attention also (Demicoli and Stichlmair 2003, Kotai et al. 2006).

The operation of different configuration is presented here for the separation of a ternary mixture where the order of decreasing volatility is A, B, C. Each configuration needs heating at the bottom (reboiler) and cooling at the top (condenser) of the equipment.

In the case of a batch rectifier (BR) the charge is filled in a heated, great volume vessel (reboiler) which is located under the column (Fig. 3.11a).

The major part of the separation is performed by the column, equipped with a condenser, where the vapor continuously arriving from the reboiler, having varying composition, is separated to distillate and liquid flowing back to the reboiler. Hence the BR can be considered as a bottom vessel column, as well. In the distillate first

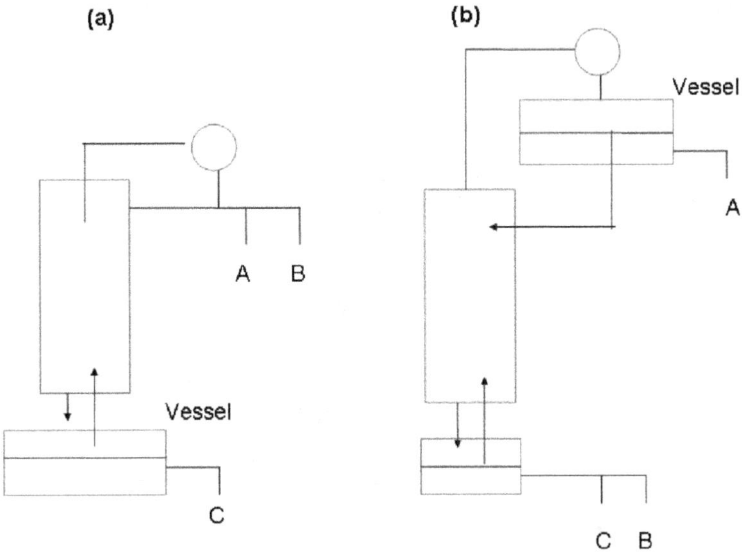

Fig. 3.11 Batch distillation configurations

A (1st cut) then B (2nd cut) can be obtained as product. C remains in the vessel and can be gained in the residue. Since in the product stream distillate, the more volatile components are enriched, each stage temperature increases with the time.

In the case of a batch stripper (BS) the vessel ensuring the feed arriving at the top stage of the column is located above the column (Fig. 3.11b). Therefore this equipment can be considered as a top vessel column. The charge is filled in this top vessel. The condensate coming from the condenser is also introduced here. The product is continuously withdrawn from the partial reboiler at the bottom of the column. So, first C (the heaviest component) then B can be obtained as product. The lightest component A remains in the vessel. Since in the product stream bottoms the less volatile components are enriched each stage temperature decreases with the time contrary to the BR.

The middle vessel column (MVC) is a combination of the BR and BS (Fig. 3.12).

Figure 3.12 shows the middle vessel column (Kotai et al. 2006).

The charge may be filled in the vessel located at the middle of the column between two column sections. From the middle vessel the liquid of varying composition is fed to the top stage of the lower (stripping) section. Both the liquid leaving the bottom stage of the upper (rectifying) section and the vapor leaving the top of the stripping section are introduced into the middle vessel. The vapor leaving the vessel arrives at the bottom of the rectifying section. The middle vessel can be considered as a stage with great, varying hold-up. Products are continuously withdrawn from both the top (A can be produced in the distillate) and the bottom (C can be produced in the bottoms).

At the end of the operation in the middle vessel B can be enriched.

Fig. 3.12 Middle vessel column

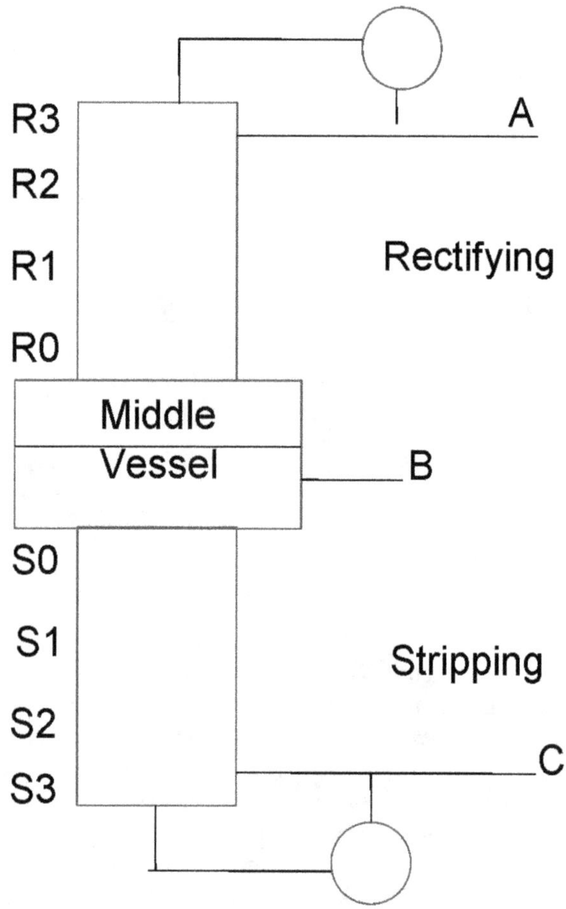

R0, R1, R2, and R3 show different levels of rectifying.

S0, S1, S2, and S3 show different levels of stripping.

The coupling of both directions, rectifying, R, and stripping, S, is of interest.

Several strategies have been proposed to increase the operating efficiency of batch distillation (Flodman and Timm 2012). Conventional batch rectification and inverted batch stripping are used cyclically to promote high product flow rates for a binary fractionation. Process controls are implemented to maintain constant product purity specifications by varying the slope of the operating line. While rectifying, the light component is removed as distillate, concentrating the heavy component in the reboiler. As a result, the distillate rate decreases with time. The process is changed from rectification to stripping modes, and the heavy component is removed as bottoms product, concentrating the light component in the distillate drum. This causes the bottoms rate to diminish with time, and the column is once again converted back

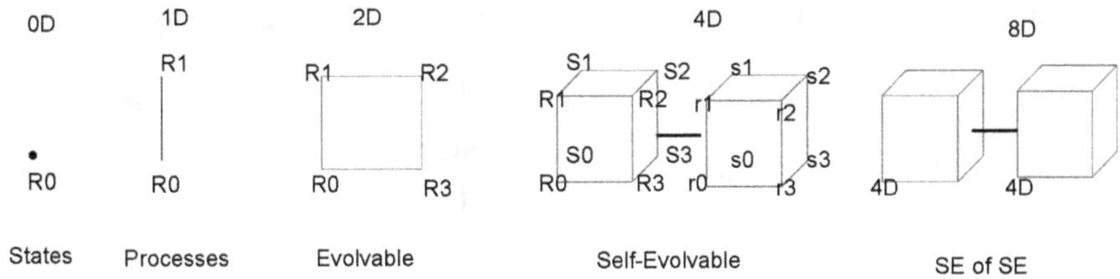

Fig. 3.13 Polytopic roadmap for middle vessel columns

to rectifying mode. Cyclic operation, transitioning from batch rectifying to stripping back to rectifying, continues until all of the initial charge is fractionated or is combined with a new charge (Flodman and Timm 2012).

The above described method illustrates the swing between constructing (rectifying) and deconstructing (stripping) steps that implies a supplementary dimension of the entire system, in this case the 4D. Correlating the constructing and deconstructing processes helps us to run separation operations that could not be done by only one of the dual batch distillation operations considered separately. The Iteration step is completed by Included Middle steps.

The process illustrated in Fig. 3.12 includes both constructing and deconstructing equipments and the middle vessel assuring and controlling the 4D polytope for processing. Middle vessel appears as a Self or IM exemplar.

This Self guides the entire separation system according to the selected criteria or rules.

Figure 3.13 shows a Polytopic Roadmap for separation

The 0D corresponds to the absence of separation. The 1D shows just one separation.

The 2D shows possible separations of different trays of the column.

It is a Rectifying, R, process.

The 4D corresponds to one device MVC-middle vessel column interconnecting two column interconnecting a Rectifying, R and a stripping S, column.

R0, R1, R2 and R3 denotes the Direct sequence, while S0, S1, S2 and S3' denotes the Reverse sequence.

We use the minuscule r0, r1, r2, r3 to denote the control of Direct sequence. We use the minuscule s0, s1, s2, s3 to denote the control of Reverse sequence.

The 4D polytope shown in Fig. 3.13 illustrates the architecture of potentially self-evolvable systems. It is a representation consisting from a pair of 3D-cubes. The second cube is a digital twin of the first cube.

The 4D-cube is obtained by joining all corners of the first 3D-cube with the corresponding corners of the second 3D-cube. The 8 lines connecting the two 3D-cubes are represented in Fig. 3.13 by a thick hyper-line. In this way the 4D-cube has a simplified representation. This representation is useful for 8D polytope too (Ammon 1998).

A thick hyper-line connects the two 7D-cubes generating the 8D polytope.

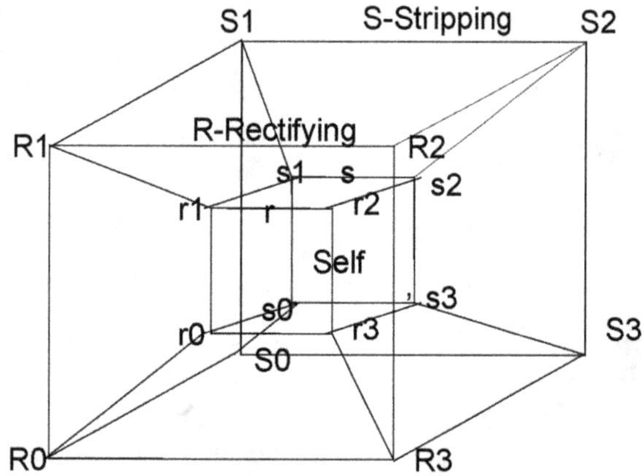

Fig. 3.14 Polytope 4D for middle vessel columns

Figure 3.14 shows more details and another presentation of the polytope 4D for middle vessel column.

Here r and s marks the activation of the corresponding faces of the inner cube. This cube corresponds to Self or in other words to Included Middle.

After any step in the connectors R-graph, the systems make contact with the Self represented by the inner cube. The inner cube is a digital twin of the outer cube.

Signals as r0, r1, r2 and r3 reinforce the evolution according the R-graph. The set (r0, r1, r2, r3) may correspond to the control of R-graph. If the Self have to change the direction shifting to s3, s2, s1, s0 the next step will be along the distributors S-graph. The set (s0, s1, s2, s3) may correspond to the control of S-graph.

Figure 3.15 shows the Polytope 8D for middle vessel columns.

It results from Fig. 3.14 by grafting.

The presented Roadmaps outlined two sequences of levels R0 → R1 → R2 → R3 that corresponds to the R levels and S3 → S2 → S1 → S0 that corresponds to the S levels.

The levels may outline a sub-level structure. The sub-levels are denoted 0, 1, 2, 3 in the direct sequence and 3’, 2’, 1’, and 0’ in reverse direction. Control frames have been considered at any level. In this way the control activities are distributed.

The 8D shows both central and distributed control capabilities.

3.4 Separations

Classification methods have been applied in chemical engineering for flow-sheets generation. Similarity based clustering method and informational criteria allow generating flow-sheets on the bases of difference in properties responsible for separation (Iordache et al. 1993a, b). For example difference in absorbability allows separation in absorption columns, difference in volatility allows separation in distillation columns, while difference in solubility allows separation in extractors.

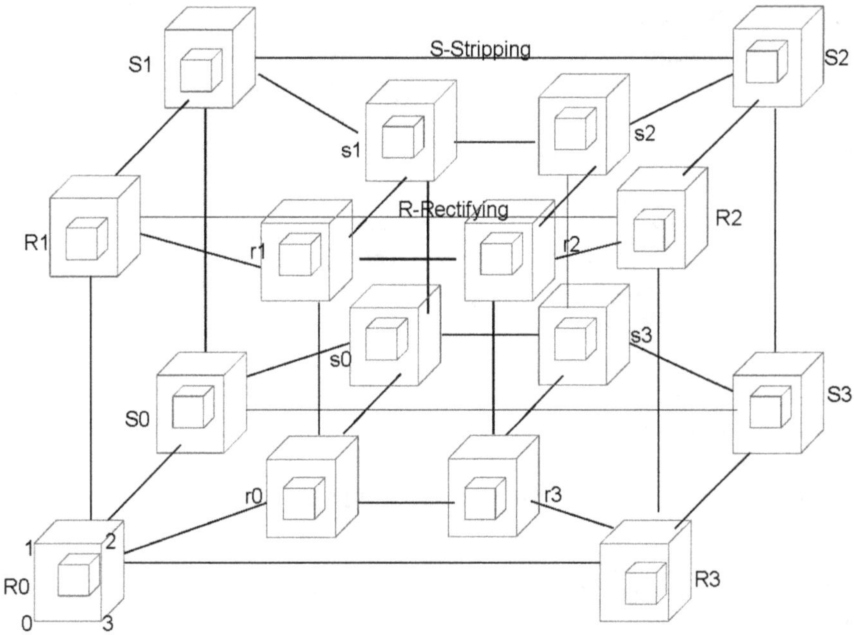

Fig. 3.15 Polytope 8D for middle vessel columns

Formal Concepts Analysis, FCA, methods can be applied for this type of classification problems. The result will be conceptual separation schemes.

Classification by FCA will be preferred to that achieved by conventional classification trees. Among the reason for this choice we note that the classification trees oriented towards compounds are too restrictive and that the tree cannot represent complex relations as "either or". Moreover, FCA supports non-disjoint classifications. In FCA the compounds are related by properties.

FCA offers a different and more complete information since it highlights the compounds separation schemes and also the properties associated to the compounds and the separation devices accomplishing the objectives. The result of a conceptual separation schemes is not a just compounds but classes of compounds with set of properties.

The focus of FCA is on formal concepts not on compounds.

The starting point for FCA application for classification and separation is a context table as Table 3.1 that contains objects or compounds to be separated and attributes or properties allowing separation.

Concept Lattices have been associated to context tables (Ganter and Wille 1999).

Table 3.1 shows an example of formal context with the set of objects or compounds,

G = {1, 2,..., 6} and the set of attributes or properties M = {a, b, c, d}. In this case the objects are chemical compounds. The properties show high values denoted

Table 3.1 Formal context for separations

Properties compounds	a	b	c	d
1	1	0	1	1
2	1	0	0	1
3	1	0	0	0
4	0	1	1	1
5	0	1	0	0
6	0	1	0	1

by "1" or low values denoted by "0". To any property, specific separation equipments may be associated.

Figure 3.16 shows the Concept Lattice for separation of six compounds based on four properties (Freeman and White 1993).

The lattice is associated to data presented in Table 3.1.

At the top of the lattice is the union of all components. The null symbol Φ indicates that at this level, no property was activated to operate classifications.

Each marked points in Concept Lattice is labeled with both the components or objects and the attributes or properties that define the separation.

To the right side of the lattice we can see that the compound 4 resulted in the light phase from the separating devices corresponding to b, d, and c as properties.

To the left side of the lattice we see that the compound 1 was separated by the devices corresponding to a, d and c as properties.

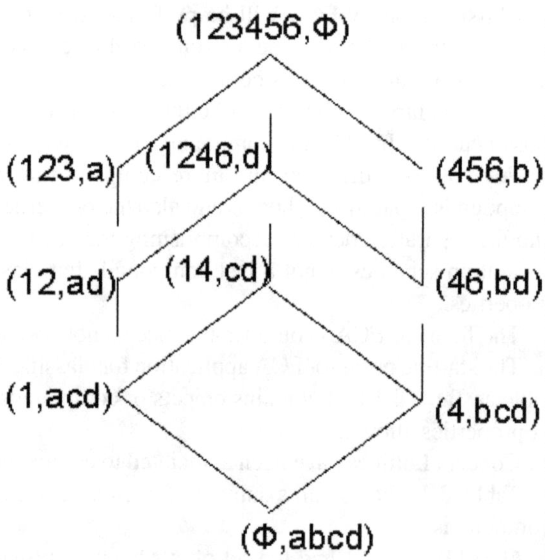

Fig. 3.16 Concept lattice for separation

As we move down, on a trajectory, we encounter smaller collections of compounds and larger collection of properties that is of separating equipments in the separation scheme.

At the bottom of the diagram we see that there are no compounds having all the properties.

Figure 3.16 outlines the intent-extent duality associated to separation.

Intent corresponds to Attributes or properties while extent corresponds to Objects or compounds.

Figures 3.17 and 3.18 shows the intent and the extent evolution during separation.

Figure 3.17 corresponds to Direct separation. It is a scheme based on a progression of Attributes that is, of separation equipments. It tracks the Fig. 3.18 in the downward direction.

The Direct separation starts from no devices and builds the different separation schemes.

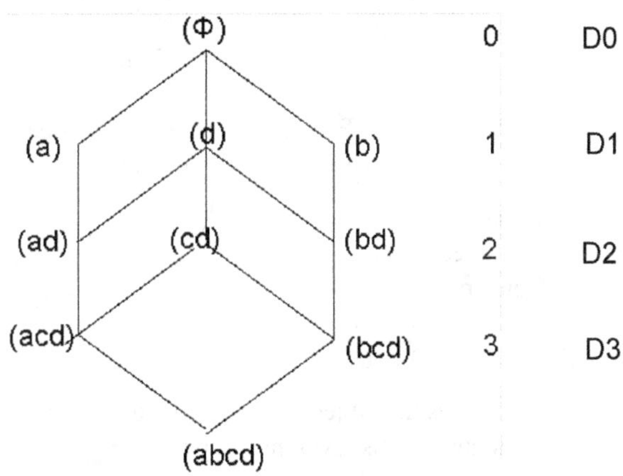

Fig. 3.17 Direct separation

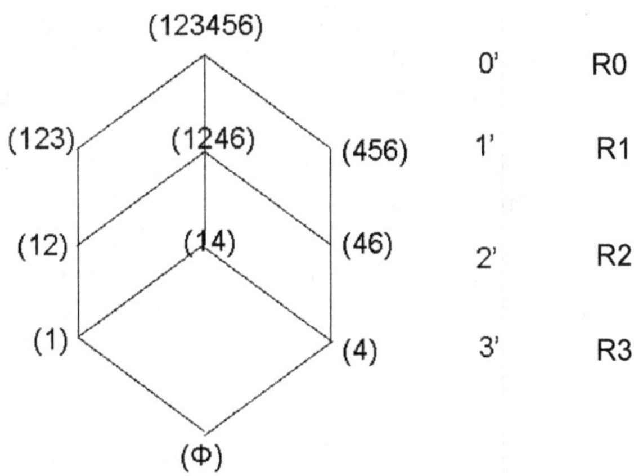

Fig. 3.18 Reverse separation

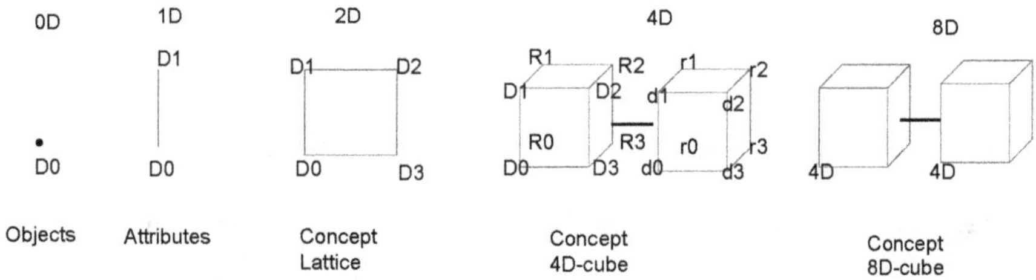

Fig. 3.19 Polytopic roadmap for separation

The levels, 0, 1, 2 and 3 shown in Fig. 3.17 corresponds to the modification in the lattice.

Figure 3.18 corresponds to Reverse separation. Is a scheme in which sequences of Objects that is of compounds develops. It tracks the Fig. 3.16 in the upward direction.

The Reverse separation starts from no Objects and builds the lattice of different mixtures.

The levels 0', 1', 2' and 3' shown in Fig. 3.18 correspond to the modification in the concept lattice.

Figure 3.19 shows a Polytopic Roadmap for separation.

The dimensions illustrated in Fig. 3.19 are:

- 0D-Objects
- 1D-Atributes
- 2D-Concept Lattices
- 4D-Concept 4D-cube
- 8D-Concept 8D-cube

The 0D shows the Objects or compounds. At the top of the lattice is the union of all components. At this level, no property was activated to operate classifications.

For any Object the Attributes are placed on a line as a 1D segment.

This may be represented as vertical column or as horizontal row.

For any Attributes the Objects may be placed on a line as a 1D segment.

Taking into account several Object Attributes lines we obtain 2D context tables and Concept Lattices. These may be tracked that is, evaluated in Direct direction, D, or Reverse direction, R, that is top-down or bottom-up in the Concept Lattice as shown in Fig. 3.19. The Direct direction refers to the Lattice of Attributes. The Reverse direction refers the Lattice of Objects.

Together, by correlation, the two directions shape a 3D frame.

The 3D frame is completed with another cube a space for dialogue between the two directions D, and R. In this way we obtain the 4D frame, the Concept 4D-cube.

D0, D1, D2 and D3 denote the Direct sequence, while R0, R1, R2 and R3 denotes the Reverse sequence.

We use the minuscule d0, d1, d2, d3 to denote the control of Direct sequence. We use the minuscule r0, r1, r2, r3 to denote the control of Reverse sequence.

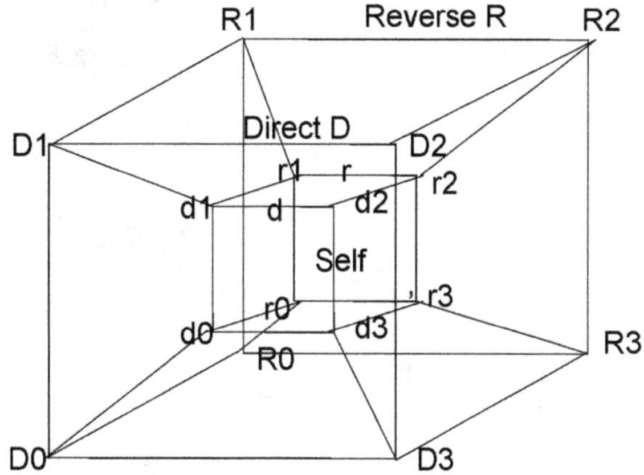

Fig. 3.20 Polytope 4D for separation

The 4D polytope shown in Fig. 3.19 illustrates the architecture of potentially self-evolvable systems. It is a representation consisting from a pair of 3D-cubes.

The second cube is a digital twin of the first cube.

The 4D-cube is obtained by joining all corners of the first 3D-cube with the corresponding corners of the second 3D-cube. It is doubling followed by relating procedure. The 8 lines connecting the two 3D-cubes are represented in Fig. 3.19 by a thick hyper-line. In this way the 4D-cube has a simplified representation. This representation is useful for 8D polytope too.

A thick hyper-line connects the two 7D-cubes generating the 8D polytope.

Figure 3.20 shows more details and another presentation of the polytope 4D for separation. It is the Concept 4D-cube. The front face corresponds to the Direct separation, that is to Attributes, while the back face corresponds to the Reverse separation, that is, to Objects. Reverse separation is in fact a combination.

Figure 3.21 shows the Polytope 8D for separation. It is the Concept 8D-cube.

It results from Fig. 3.20 by grafting.

After any step the inner cube the Self is involved and direct the next step in separation. The control levels are indicated on the inner cube.

The inner cube is a digital twin of the outer cube.

The presented Roadmaps outlined two sequences of levels D0 → D1 → D2 → D3 that corresponds to the D levels and R3 → R2 → R1 → R0 that corresponds to the R levels.

The levels may have a sub-level structure. The sub-levels are denoted 0, 1, 2, 3 in the direct sequence and 3', 2', 1', and 0' in reverse direction. Control frames have been considered at any level. In this way the control activities are distributed.

The 8D shows central and distributed control capabilities.

Observe that the presented method is an attempt to unify the design, scheduling and control for separation problems. Burnak et al. (2019) discuss other facets of this unification.

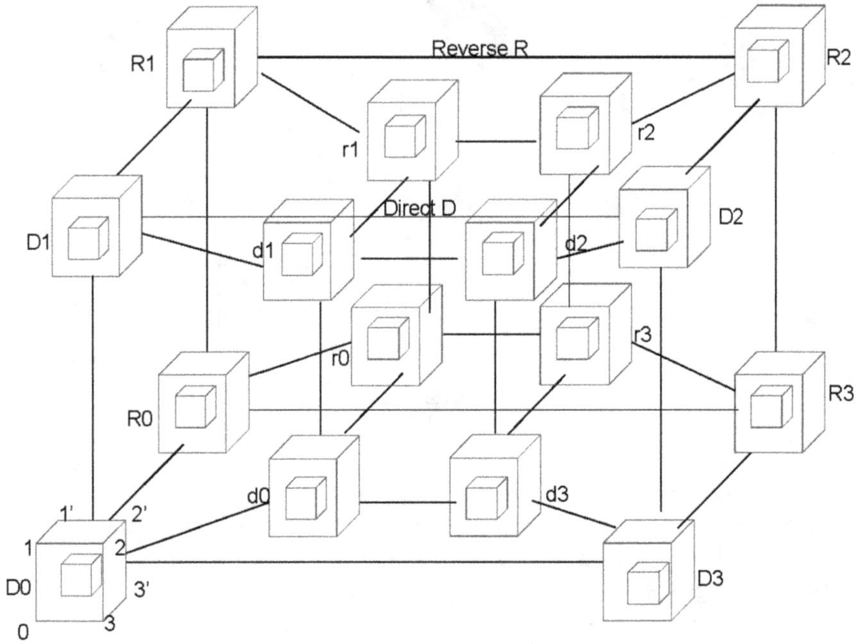

Fig. 3.21 Polytope 8D for separation

References

Ammon, J.: Hypercube Connectivity within ccNUMA Architectures. SGI Origin Team, Mountain View, CA (1998)

Burnak, B., Diangelakis, N.A., Pistikopoulos, E.N.: Towards the grand unification of process design, scheduling, and control–Utopia or reality? Processes **7**(7), 461 (2019)

Coppens, M.O.: A nature-inspired approach to reactor and catalysis engineering. Curr. Opin. Chem. Eng. **1**(3), 281–289 (2012)

Demicoli, D., Stichlmair, J.: Novel operational strategy for the separation of ternary mixtures via cyclic operation of a batch distillation column with side withdrawal. Chem. Eng. Trans. **3**, 361–366 (2003)

Ehrenfeld, J.R.: Sustainability by Design: A Subversive Strategy for Transforming our Consumer Culture. Yale University Press, New Haven, CT (2008)

Ehrenfeld, J.R., Hoffman, A.J.: Flourishing: A Frank Conversation About Sustainability. Stanford University Press, Stanford, CA (2013)

Flodman, H.R., Timm, D.C.: Batch distillation employing cyclic rectification and stripping operations. ISA Trans. **51**(3), 454–460 (2012)

Fomin, S.: Dual graphs and Schensted correspondences. In: Leroux, P., Reutenauer, C. (eds.) Series formelles et combinatoire algebrique, pp 221–236. Montreal, LACIM, UQAM (1992)

Fomin, S.: Duality of graded graphs. J. Algebr. Combin. **3**, 357–404 (1994)

Freeman, L., White, D.: Using Galois lattices to represent network data. Soc. Netw. **23**, 127–146 (1993)

Ganter, B., Wille, R.: Formal Concept Analysis. Mathematical Foundations. Springer, Berlin (1999)

Iordache, O., Corriou, J.P., Garrido-Sanchez, L., Fonteix, C., Tondeur, D.: Neural network frames applied for biochemical kinetic diagnosis. Comp. Chem. Eng. **17**, 1101–1113 (1993a)

Iordache, O., Corriou, J.P., Tondeur, D.: Separation sequencing. Use of information distance. Can. J. Chem. Eng. **71**(12), 955–966 (1993b)

Kotai, B., Lang, P., Balazs, T.: Separation of maximum azeotropes in a middle vessel column. In: Institution of Chemical Engineers Symposium Series, vol. 152, p. 699–708. Institution of Chemical Engineers, 1999 (2006)

Kritzinger, W., Traar, G., Henjes, J., Sihn, W., Karner, M.: Digital twin in manufacturing: a categorical literature review and classification. IFAC-Papers On Line **51**(11), 1016–1022 (2018)

Thiel, T.: The design of the connection machine. Des. Issues **10**, 1, 5–18 (1994)

Chapter 4
Modeling

Abstract Modeling and simulation pertains to the intelligence level of reality. So, modeling methods will adopt hierarchical reference architectures. The dimensions of the Polytopic Roadmap for modeling include: Data, Model, Simulation, Digital Twin (DT) and DT of DT. The simulation level includes: Data, Model, Simulation Tools and System Design. The frames supporting live-like, 4D, and intelligent-like, 8D, stages are illustrated here. Relation between 4D frame and industry 4.0 is emphasized. The model based engineering, MBE, is a reference architecture presented as an 8D framework in development.

4.1 Digital Twins

Modeling and simulation pertains to the intelligence level of reality (Hartmann 1952).

Since the reality may be structured in levels and sub-levels, it is expected that the modeling methods will adopt hierarchical reference architectures with levels and sub-levels (Poli 2001). Moreover, it is expected that the roadmap to complexity for reality parallels the roadmap for modeling or in other words for virtuality (Iordache 2021, 2022).

Figure 4.1 shows a Polytopic Roadmap for simulation and modeling.

It outlines the development stages for modeling and simulation (Rosen et al. 2015).

The dimensions of the Polytopic Roadmap as illustrated in Fig. 4.1 are:

- 0D-Data-D
- 1D-Model-M
- •2D-Simulation
- 4D-Digital Twins-DT
- 8D-DT of DT

The Simulation level includes:

- K0-Data-D
- K1-Model-M
- K2-Simulation Tools-ST

O. Iordache, *General Reference Architecture Frameworks*, Studies in Systems, Decision and Control 73, https://doi.org/10.1007/978-3-031-70718-6_4

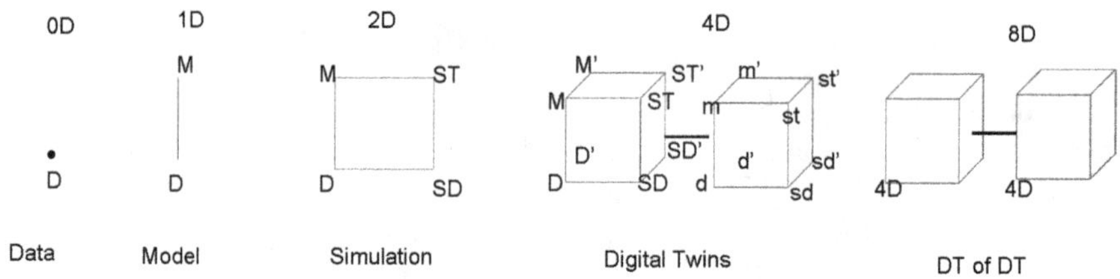

Fig. 4.1 Polytopic roadmap for modeling

- K3-System Design-SD

We use apostrophes 'to denote the modified dual frames and minuscule to denote the associated control values.

So we have:

- K0'-Data'-D'
- K1'-Model'-M'
- K2'-Simulation Tools'-ST'
- K3'-System Design'-SD'
- k0- Data control-d
- k1-Model control-m
- k2-Simulation Tools control-st
- k3-System Design control'–sd
- k0'- Data control'-d'
- k1'-Model control'-m'
- k2'-Simulation Tools control'-st'
- k3'-System Design control'-sd'

The 0D in Fig. 4.1 shows Data, D.

The 1D shows Data, D and the individual application Model M. The 2D frame called Simulation is more developed. It includes D, M, Simulation Tools, ST and System Design, SD.

The 0D corresponds to Data about objects or states before any modeling.

The 1D corresponds to individual applications model. Modeling is limited to specific topics by experts in very specific topics as for instance mixing, chemical kinetics and so on. Experts use this kind of models to understand and predict phenomena, to perform analysis, to design and schedule. This stage is represented in Fig. 4.1 as the 1D segment suggesting that at this level the experts are able to select the Data represented as a dot and to predict by modeling the 1D line to be followed.

The 2D Simulation includes also the simulation tools, ST. Standard simulation tools are developed for specific domains as for instance, statistical analysis, fluid mechanics, flow-sheeting and so on. The 2D representation suggests that the simulation tools are able to run several 1D models Application to connect them in a ST frame and this in turn included in an SD frame. Computer aided design, CAD, and computer aided product design and computer aided process engineering, CAPE may be classified here. The simulation tools allow design validation, optimization and decision support.

Simulation allows a systemic approach to multi-level and multi-disciplinary systems with enhanced range of applications as for instance model based engineering, MBE, and model based systems engineering, MBSE, (Estefan 2007; Rosen et al. 2015).

The roadmap from Fig. 4.1 outlines the strategy to follow after the 2D stage.

The 2D level capabilities are surpassed by Digital Twin, DT, associated to 4D.

These increase dimensionality to 4D and close the loop making possible a dialogue between Simulation (D → M → ST → SD) and Resimulation (SD' → ST' → M' → D').

The resulting 4D framework is considered as an essential module for self-evolvable, that is sustainable, reference architectures.

The 4D shows a cube of simulation values and another cube of control values.

The control cube is a digital twin of the simulation values cube.

The thick line shown in Fig. 4.1 is a simplified presentation of the relations between the two cubes from the 4D frame (Thiel 1994).

The relation between 4D frame and industry 4.0 is significant.

The fourth industrial revolution that is Industry 4.0, driven by the internet and advancing in several enabling technologies, promises a networked world of smart systems, products, procedures and processes (Schwab 2016). It is this vision that has led to increased interest in the Digital Twin, DT concept—the virtual representation of real-world entities to facilitate integration of real-world with digital systems.

The Industry 4.0 vision, and specifically the development of Digital Twins, has made an impact in social and economical domains.

Digitalization becomes part of industrial processes and opens up new application areas. The vision of Industry 4.0 as summarized by Kasper et al. (2020), spots a picture of a new level of dialoguing and interconnectedness of devices and smart services in the Internet of Things. Systems and products will be created that have a high degree of networking, flexibility and automation. Due to these features, they have a potential for reducing cost and energy consumption (Taylor et al. 2019; Hartmann and Van der Auweraer 2021).

An early definition for the Digital Twin, as proposed by Glaessgen and Stargel, (2012), describes it as a multi-physics, multi-scale, and probabilistic simulation of a complex product that uses models, sensor updates, and so on, to mirror the life of its corresponding physical twin. This definition has been subsequently understood and applied in different ways. This is reflected in the differentiation made between digital models, digital shadows and Digital Twins in the manufacturing domain, (Rosen et al. 2015; Kritzinger et al. 2018; Tao et al. 2019; Jones et al. 2020).

For the manufacturing domain, Siemens has advocated that Digital Twin developments are the key enabler of autonomous production systems of the future (Rosen et al. 2015).

Digital Twins should play an enabling role by gathering information in each stage of the product lifecycle and making it available for dialogue to subsequent stages in the system.

With Digital Twins, organizations can not only create mirrors of real-world objects and processes but integrate physical and virtual worlds through bidirectional flows of information. With real-time simulations and intelligent algorithms, Digital Twins shifts the focus of data-driven operations from post monitoring to ante predictions and optimization in increasingly complex environments. Digital Twins are said to revolutionize the manufacturing industry, but it may also have major impacts on future studies and the general ways organizations anticipate the future. A Digital Twin development can be a new advanced form of project planning.

The Polytopic Roadmap shown in Fig. 4.1 illustrates the Lupasco principle of antagonistic dualism (Lupasco 1951). It outlines the principle of antagonistic dualism in the step by step transition from 0 to 8D.

As shown in Fig. 4.1 the inclusive DT represents a 4D frame if the connections from Simulation, s, and Resimulation, r, worlds are structured by the digital control cube, in 4D. The control cube is a digital twin of the simulation cube.

.The 8D stages, identified here as DT of DT, results by applying the grafting strategy to the 4D stage. This means replacing the vertices in the initial 4D by 4D frameworks. It results a 4D of 4D frames.

Can be achieved looking to the detailed structure of the vertices in 4D. It is possible if the 4D level is generalized. The intelligence level by grafting was defined as the possibility that life will flourish forever (Ehrenfeld 2012; Ehrenfeld and Hoffman 2013).

4.2 Frameworks

Figure 4.2 shows more details and another presentation of the polytope 4D for modeling.

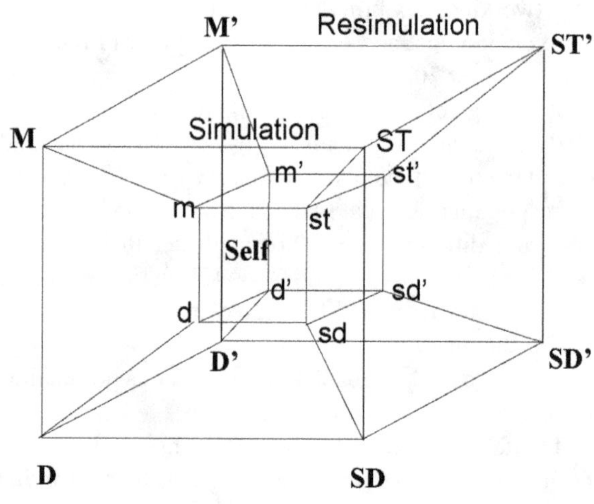

Fig. 4.2 Polytope 4D for modeling

Figure 4.3 shows a developed roadmap of polytope 4D for modeling.

The 2D and 4D frames are presented in a developed form here.

Simulation sequence is considered as 2D frame.

Simulation is correlated to Resimulation. To accomplish this, a new frame called the Self or Included Middle should be considered.

The Self is shown with Simulation, s, part and Resimulation r, part. Since Simulation and Resimulation frames are 2D the s and r frames should be 2D too.

Relating s and r allows obtaining an inner cube. Finally the 4D frame will result, by inclusion the inner cube in the outer cube. It will be a 4D-cube representing the self-evolvable systems. The inner cube is a digital twin of the outer cube (Fig. 4.2).

Figure 4.4 shows the polytope 8D for modeling.

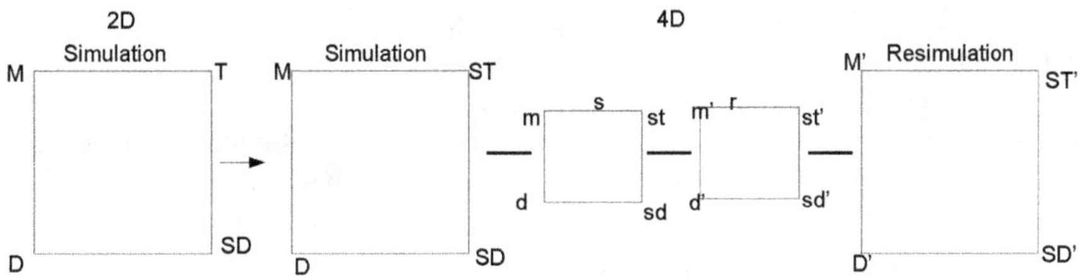

Fig. 4.3 Developed roadmap for polytope 4D for modeling

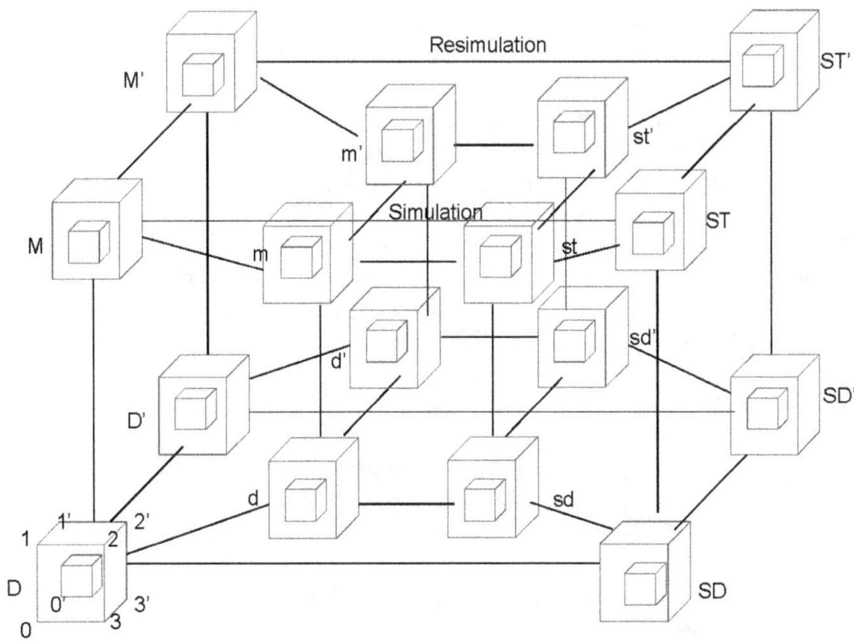

Fig. 4.4 Polytope 8D for modeling

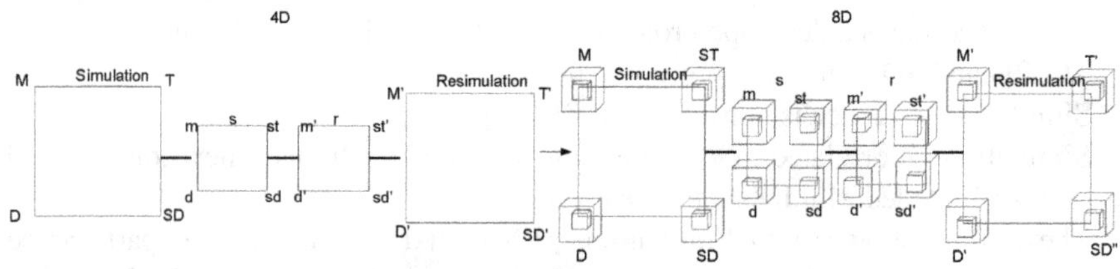

Fig. 4.5 Developed roadmap for polytope 8D for modeling

This is the GRAF dedicated to 8D Program.

Figure 4.4 is obtained from Fig. 4.2 by grafting strategy.

The sub-levels in D may be denoted by 0, 1, 2, 3 for a direct sub-sequence in D, and by 3', 2', 1', 0' for a reverse sub-sequence in D'. For a complete description of D, the sub-self of D components should be identified.

The presented Roadmaps outlined two sequences for evolution the direct sequence: D → M → ST → SD that corresponds to the Simulation stages and the reverse sequence: SD' → ST' → M' → D' that corresponds to the Resimulation stages.

The directly acquired development is questioned and modified by the reverse sequence. The bottom-up construction from direct sequence complements the top-down deconstruction of the reverse sequence.

Figure 4.5 shows a developed roadmap for polytope 8D for modeling.

The 4D and 8D frames are presented here.

The initial 4D architecture shows the developed frames of external and internal cubes. These cubes relate the corresponding 2D frames.

The transition from 4 to 8D is achieved by replacing any 0D vertex from the initial 4D architecture by component 4D frames. It is a grafting strategy that keeps the initial 4D architecture but replace any 0D vertices by component 4D vertices.

It results a 4D hypercube of 4D hypercubes. The new dimension is 8D = 4D + 4D. In this sum, the first 4D come from the initial 4D frame and the next 4D from the component 4D frames replacing the 0D vertices of the initial 4D hypercube

In the next step an inclusion of a new frame called the Included Middle or Self is considered. The Self is shown with Simulation s, part and Resimulation r, part.

Relating s and r allows obtaining the inner 7D-Cube. Finally the 8D frame results, by inclusion of the inner 7D-cube in the outer 7D-Cube. The inner 7D-cube is a digital twin of the outer 7D-cube.

The development roadmap may be interpreted in terms of dual process theory (Kahneman 2011). The Simulation sequence D → M → ST → SD pertains to System 1 while the Resimulation sequence SD' → ST' → M' → D' pertains to System 2.

A theory in which the control system (System 3, S3) takes precedence was developed (Houdé 2019). S3 is associated to the Self putting together the r and s parts of Included Middle.

4.3 Model Based Engineering

MBE, model based engineering, may be considered as a significant step towards 8D framework for modeling. It starts with a 4D polytope.

To replace the linear thinking associated with the V-models, some companies encourages engineers to adopt MBSE, model based systems engineering, to arrive at a faster and more agile system development process (Estefan 2007).

Stacks of paper documents are being replaced with digital twins and digital thread.

For its part, company Boeing has developed the so-called "Boeing diamond" as a new way to envision the systems engineering process and illustrate the addition of real-time modeling, simulation, and collaboration to the traditional V-model (Seal et al. 2018).

Figure 4.6 shows a preliminary form of the 4D polytope for MBE.

The bottom portion includes activities related to the physical system as design and delivery. This portion closely resembles the conventional V-models diagram. The top portion includes activities correlated to the digital twin of the physical system and includes detailed and real-time modeling and simulation. In between is the digital thread that is available to all design teams, including mechanical, electric, electronic, software, and so on, and provides real-time feedback throughout the design, delivery, and deployment of the system. This diamond illustrates the overlay of digital twins

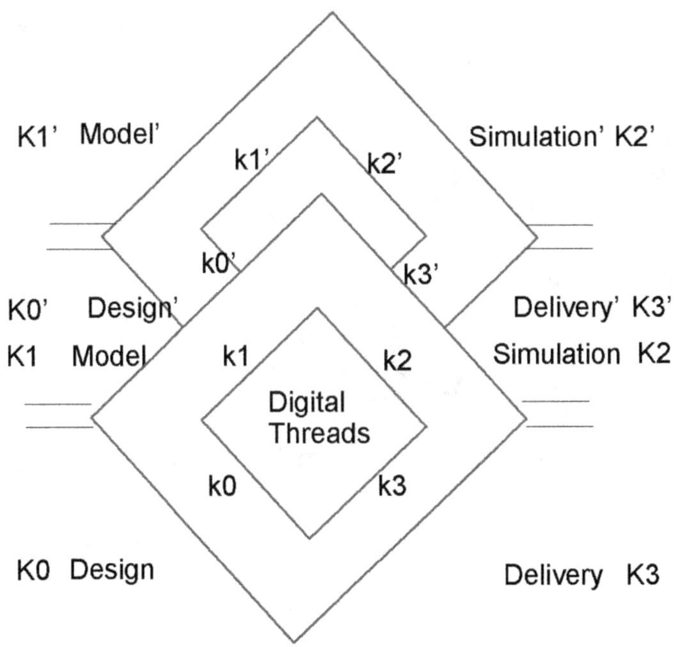

Fig. 4.6 Polytope 4D for MBE-preliminary form

on the traditional V-models and the collaborative and iterative nature of MBSE as supported by the digital thread.

Notations are:

- K0-Design-D
- K1-Model-M
- K2-Simulation-S
- K3-Delivery-Del
- K0'-Design'-D'
- K1'-Model'-M'
- K2'-Simulation'-S'
- K3'-Delivery'-Del'

We make use of minuscule to denote the control level, included in Digital Thread.

The notations are:

- k0- Design control-d
- k1- Model control-m
- k2- Simulation control-s
- k3- Delivery control-del

The duality or modifications are labeled by apostrophes '.

Figure 4.7 shows a standard form of the 4D polytope for MBE.

The sub-levels should be considered in the next step of complexification.

We may replace the basic levels D, M, S, Del by more detailed frames.

At the limit these grafting frames will be 4D.

For D-Design we denote the sub-levels:

- 0-planned
- 1-designed
- 2-specified

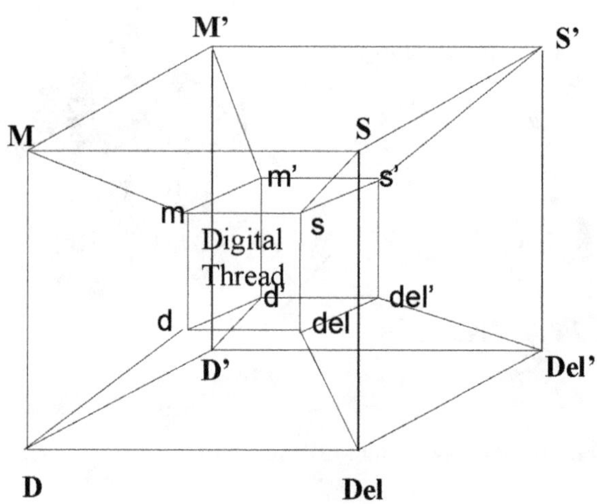

Fig. 4.7 Polytope 4D for MBE

- 3-offered
- 4-needed

For M-Model we denote the sub-levels:

- 0-planned
- 1-definition
- 2-system engineering
- 3-bussiness
- 4-market

For S-Simulation we denote the sub-levels:

- 0-production
- 1-qualification
- 2-certification
- 3-operation
- 4-ecosystem

For Del-Delivery we denote the sub-levels:

- 0-built
- 1-tested
- 2-certified
- 3-delivered
- 4-supported

Figure 4.8 shows a preliminary form of a Polytope 8D for MBE.
Figure 4.9 shows a standard form of a Polytope 8D for MBE.

4.4 Roads

Figure 4.10 shows the roads to higher dimensional modeling:

S1 and S2 roads have been outlined. S2 is based on Aristotelian logic while S1 support a non-Aristotelian logic.

Consider that the S2 road starts from Data, 0D, represented here as a circle. Included in this an S1 frame appears. It is represented as an included dot. The dot and the circle will be related in a segment or line, 1D, as shown on the S2 road. The segment corresponds to Model.

Then on the S1 road an included middle represented as a segment appears. Transition to S2 road imposes the development of the segments. By relating the two segments the 2D square results.

Transition to S1 road imposes the development of two squares. The squares are oriented. By relating these squares an unrepresented cube may result. On the S1 road an included cube appears and develops. Transition to S2 road imposes the

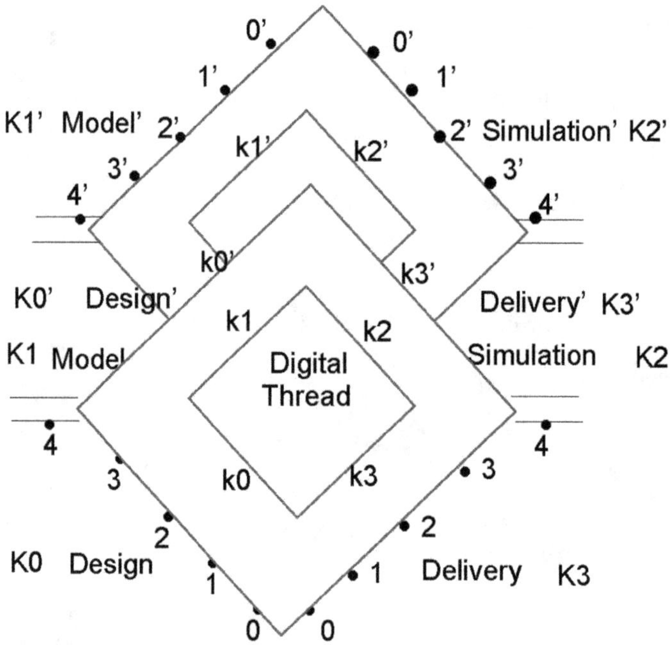

Fig. 4.8 Polytope 8D for MBE-preliminary form

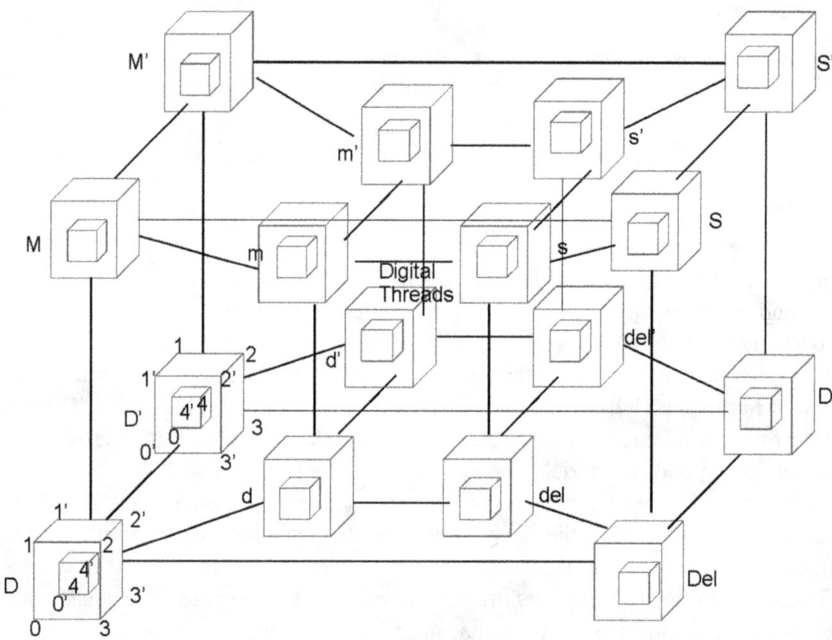

Fig. 4.9 Polytope 8D for MBE

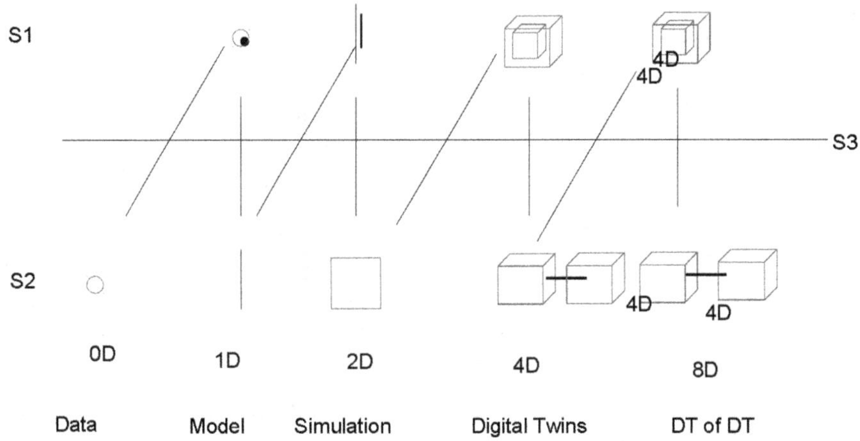

Fig. 4.10 Roads to higher dimensional modeling

development and the correlation of the two cubes. The 4D cube results. It is associated to Digital Twins, DT.

The 4D-cube on the S2 road is obtained by joining all corners of the first 3D-cube with the corresponding corners of the second 3D-cube. The 8 lines connecting the two 3D-cubes are represented in Fig. 4.10 by a thick hyper-line. In this way the 4D-cube has a simplified representation. This representation is useful for 8D polytope too.

The dimensions of stages shown in Fig. 4.10 are:

- 0-Data
- 1D-Model
- 2D-Simulation
- 4D-Digital Twins
- 8D-DT of DT

The framework 8D is resulting as 4D of 4D architecture.

On the S1 road an included hypercube appears and develops. Transition to S2 road imposes the development and the correlation of two hypercubes. Now, the dimension is 8D = 4D + 4D. The 8D-cube results. This represents the DT of DT stage.

The process continues indefinitely in the same way as a zigzag between S1 and S2 roads.

References

Ehrenfeld, J.R.: Beyond the brave new world: Business for sustainability. In: Bansal, P., Hoffman, A.J. (eds.) The Oxford Handbook of Business and the Natural Environment, pp. 611–619. Oxford University Press, Oxford (2012)

Ehrenfeld, J.R., Hoffman, A.J.: Flourishing: A Frank Conversation About Sustainability. Stanford University Press, Stanford, CA (2013)

Estefan, J.A.: Survey of model-based systems engineering (MBSE) methodologies. Incose MBSE Focus Group **25**(8) (2007)

Glaessgen, E., Stargel, D.: The digital twin paradigm for future NASA and U.S. Air Force vehicles. In: Structures, Structural Dynamics, and Materials Conference, pp. 1–14 (2012)

Hartmann, D., Van der Auweraer, H.: Digital twins. In: Progress in Industrial Mathematics: Success Stories, pp. 3–17. Springer, Cham (2021)

Hartmann, N.: The New Ways of Ontology. Greenwood Press, Westport (1952)

Houdé, O.: 3-System Theory of the Cognitive Brain, A Post-Piagetian Approach to Cognitive Development. Routledge, London (2019)

Iordache, O.: Polytopic Roadmaps. Springer, Cham (2021)

Iordache, O.: Roads to Higher Dimensional Polytopic Projects-Reference Architectures. Springer, Cham (2022)

Jones, D., Snider, C., Nassehi, A., Yon, J., Hicks, B.: Characterising the digital twin: a systematic literature review. CIRP J. Manuf. Sci. Technol. **29**, 36–52 (2020)

Kahneman, D.: Thinking, Fast and Slow. Farrar, Straus and Giroux, New York (2011)

Kasper, L., Bacher, T., Birkelbach, F., Hofmann, R.: Digitalization possibilities and the potential of the digital twin for steam supply systems. VGB Power Tech **11**, 45–52 (2020)

Kritzinger, W., Traar, G., Henjes, J., Sihn, W., Karner, M.: Digital twin in manufacturing: a categorical literature review and classification. IFAC-Papers On Line **51**(11), 1016–1022 (2018)

Lupasco, S.: Le principe d'antagonisme et la logique de l'énergie. Prolégomènes à une science de la contradiction. Hermann & Co., Paris (1951)

Poli, R.: The basic problem of the theory of levels of reality. Axiomathes **12**, 261–283 (2001)

Rosen, R., Von Wichert, G., Lo, G., Bettenhausen, K.D.: About the importance of autonomy and digital twins for the future of manufacturing. IFAC-Papers On Line **28**(3), 567–572 (2015)

Schwab, K.: The fourth industrial revolution. World Economic Forum, Geneva (2016)

Seal, D., Farr, D., Hatakeyama, J., Haase, S.: The system engineering VEE – is it still relevant in the digital age? In: Proceedings of the 2018 National Institute of Standards and Technology (NIST) Model Based Enterprise Summit (2018)

Tao, F., Qi, Q., Wang, L., Nee, A.Y.C.: Digital twins and cyber–physical systems toward smart manufacturing and industry 4.0: correlation and comparison. Engineering **5**(4), 653–661 (2019)

Taylor, N., Human, C., Kruger, K., Bekker, A., Basson, A.: Comparison of digital twin development in manufacturing and maritime domains. In: International Workshop on Service Orientation in Holonic and Multi-Agent Manufacturing, pp. 158–170. Springer, Cham (2019)

Thiel, T.: The design of the connection machine. Des. Issues **10**, 1, 5–18 (1994)

Chapter 5
Design

Abstract The creative and the analytical ingredients in engineering education should stand out as complementary. Both analytical and intuitive methods are required to improve engineering design. Dual process designs, and processes integration are presented here in the general frame for complexity. The design roadmaps are based on the wave equation solution in different fields. Galois fields offer creative designs models, while real field provides analytic designs. Integration shows the wave equation as source for designs of experiments. It is also applied in the study of informational entropy production for distillation columns. Divergence and convergence methodology in design is presented here as a reference architecture framework.

5.1 Dual Process Design

Dual process theory was widely accepted as a basis for understanding human cognition and behavior (Kahneman 2011; Evans and Stanovich 2013). However its potential to bring clarity to design methodology has only recently realized and applied (Moore et al. 2014, Cash et al. 2019).

The standard engineering education primarily teaches students to use analytical methods when solving problems, which are effective in most real-world situations. On the other hand, heavily analytical approaches, as for instance partial differential equations, PDE, may hinder creative output. Therefore more intuitive methods are required to increase novelty part in engineering design.

Dual process theory is an established model in psychology and human decision making that separates fast, intuitive, heuristic or System 1 processes, from slow, analytical, formal or System 2 processes (Kahneman 2003, 2011; Croskerry 2009; Overton and Ricco 2010).

To this point, the theory has been rarely applied to design methodology (Moore et al. 2014). Preliminary results suggest that System 1 intuitive thinking is correlated with a higher average idea novelty up to a threshold.

O. Iordache, *General Reference Architecture Frameworks*, Studies in Systems, Decision and Control 73, https://doi.org/10.1007/978-3-031-70718-6_5

A balance of System 1, S1, and System 2, S2, thinking could increase novelty potential and design effectiveness. This balance may be under the control of the System 3, S3.

Lupasco (1951), Houdé (2019) go beyond the approaches of Kahneman (emphasizing System 1, S1) and Piaget (emphasizing System 2, S2) to put forward a theory where the included third T, or the inhibitory-control system (System 3, S3) takes precedence.

T or S3 is the bridge between S1 and S2. It is supposed that the brain support S3 as a third control system located in the prefrontal cortex which is dedicated to inhibiting Kahneman's intuitive heuristics System 1 and activating Piaget's logical algorithms System 2 anywhere in the brain on a case-by-case basis, depending on the goal and context of the problem to solve.

Understanding this relationship and the importance of intuitive thinking in the design process is important to improve the value of conceptual design thinking and may have implications in design education and modeling cognitive design processes.

The impact of intuitive thinking in design idea generation and the ways in which intuitive thinking can be effectively included in design education and design methods start to be investigated. Taking a dual process approach, it was considered that design ideas are generated through a process composed of S1, that is, intuitive thinking and S2 that is, analytical thinking and both quality and quantity of design ideas generated may depend on how different types of thinking have been applied.

A case study in which both type of thinking S1 and S2 may be supported by the same formal mathematical model is presented in the following.

The model describing both S1 as S2 approaches is a first order wave equation, WE (Iordache 1996, 2010).

If the model parameters pertain to finite structures as Galois Fields, GF, the solutions of the WE are designs of experiments, DoE (Iordache 1996, 2010, 2021).

These DoE represent in fact embedded technological scheme. They are the main result of intuitive thinking, S1.

If the model parameters are defined on the real field, the WE is the usual PDE describing the transfer phenomena in the equipment to be designed. The PDE pertain to analytic thinking S2.

The proposed modeling strategy is to keep the pattern of the real models but to adapt the definitions of parameters, functions and operations to specific features of the studied phenomena. In this way we have universal structurally stable models admitting some formulations invariant under the choice of the algebraic structure of numbers for space, time and other parameters. The models and methods for the intuitive design of complex systems appear as dual of the models and methods based on conventional real field.

The point of view adopted for high complexity systems modeling is that the algebraic frame for variables as time, space and parameters must agree first of all with the specific context of phenomena and analysis of the studied system.

Replacing universal concepts for time and space, by specific relational concepts as, multi-dimensional, multi-scaled, dyadic, and cyclic time and space is permitted and should be operated if the system empirical analysis can proceed on this basis (Gould 1987, Iordache 1992, 1996, 2010, Overton 1994).

The key message is that we should think about the future designs as high dimensional.

When designing the vision, and the paths towards that vision, we are not looking towards a singular end-path. Rather, we are opening up space and time to fulfill many simultaneous objectives. So, we need more than Newtonian space and time, because future systems are not only reliant on physiological physical dimensions, but also include augmented and virtual dimensions of thinking. When we think of the future design the mental map we need is not only 3D-cube but also includes the context conditions.

The implementation of visionary high dimensional thinking requires pursuing stacks of the 3D-cube framework as a starting point of forward-looking design and business modeling (Grothendieck 1983).

The reformulation of the models with multi-scaled or cyclic parameter is based on experimental observation of systems and may be used to redesign other systems in order to become multi-scaled, cyclic and so on (Iordache et al. 1990, 1993).

A case study concerning pressure swing adsorption, PSA, will be considered more in detail here (Ruthven 1984; Ruthven et al. 1994; Tondeur et al. 1985).

Cycling operation methods are of great importance in oil chemistry, in pharmaceutical and food industry, bio-refinery, isotopes separation, hydrogen purification, desalinization, and so forth. Cyclic separation technologies such as pressure swing adsorption, PSA, are unsteady non-linear processes difficult to put into practice and to control.

In PSA processes, gas mixtures are separated by selective adsorption over a bed of sorbent materials. The cyclic nature of these processes arises from the succession of high pressure adsorption phase and the subsequent low-pressure regeneration phase.

For analytical design of PSA processes we refer to the first order wave equation representing the material balance for transport processes. The real field wave model shows that the variation of concentration, in time and space, is due to convective processes and to kinetic processes of interaction.

The model (5.1) describes phenomena of major physical and technological interest, as convective momentum, heat and mass transfer accompanied by chemical reaction.

$$\frac{\partial y}{\partial t} + v\frac{\partial y}{\partial z} + q(y) = 0 \tag{5.1}$$

Here y (t, z) denote the concentration variable in t-time and z-space, v denotes the drift velocity, and q(y) denotes the interaction rate (Ruthven 1984; Ruthven et al. 1994; Rhee et al. 1989).

PSA systems are distributed in nature with temporal and spatial variations that are represented by PDE systems. These include equilibrium isotherms, mass and energy balances, as well as detailed thermodynamic and transport phenomena models.

Table 5.1 Sum in GF(2)

⊕	0	1
0	0	1
1	1	0

All the parameters of the models as (5.1) are defined over the mathematical real field.

This field may grow to be unsuitable for control in real time of high complexity systems.

In such cases, the parameters and the functions of the real field conventional PDE models for transfer have to be replaced by different algebraic structures to take into account from start the specificity of that system. The finitude and cyclicity are specific for PSA.

For S1 model we will use GF(m) field where m is the number of elements of the field.

This field was successful to describe the finite, logical or cyclic type of processes (Lidl and Niederreiter 1997).

The sum operations in GF(2) is presented in Table 5.1.

The sum operations in GF(4) is shown in Table 5.2.

The sum operations in GF(8) is shown in Table 5.3.

The wave equation, WE, equation useful for intuitive design of complex systems will be introduced in what follows (Iordache 1996, 2021).

Table 5.2 Sum in GF(4)

⊕	0	1	2	3
0	0	1	2	3
1	1	0	3	2
2	2	3	0	1
3	3	2	1	0

Table 5.3 Sum in GF(8)

⊕	0	1	2	3	4	5	6	7
0	0	1	2	3	4	5	6	7
1	1	0	3	2	5	4	7	6
2	2	3	0	1	6	7	4	5
3	3	2	1	0	7	6	5	4
4	4	5	6	7	0	1	2	3
5	5	4	7	6	1	0	3	2
6	6	7	4	5	2	3	0	1
7	7	6	5	4	3	2	1	0

We will use the space Z interpreted as beds or operation space, the time T interpreted as the separation step, output Y, interaction rate Q and velocity V.

One of the simplest mechanisms of separation is that in which small changes of the separation step, ∂T, are associated to small changes of the output, ∂Y. Moreover, the change of output ∂Y depends on the existing output Y and the change ∂T of T that is:

$$\frac{\partial \mathrm{Y}}{\partial \mathrm{T}} \propto \mathrm{Q(Y)} \tag{5.2}$$

This mechanism is that of kinetic type.

Another separation mechanism, takes into account that the variation of the output Y, along the separation steps T is proportional to the output variation along the beds or operations space Z. Separation, means in fact travel in time T, along the space of operations, Z. As Z is screened with a velocity V, the output varies proportionally.

This means that:

$$\frac{\partial Y}{\partial T} \propto V \otimes \frac{\partial Y}{\partial Z} \tag{5.3}$$

Here the velocity is a vector V pertaining to the same algebraic structure as Z, T, Q or Y.

This mechanism is that of convection or drift type.

The general model of the separation process including both types of separation processes, corresponding to the evolution according to T and Z, is the first order wave equation WE:

$$\frac{\partial Y}{\partial T} \oplus V \otimes \frac{\partial Y}{\partial Z} \oplus Q(V) = 0 \tag{5.4}$$

The initial condition is:

$$\mathrm{Y(Z, 0)} = \mathrm{F(Z)} \tag{5.5}$$

The first-order wave equation, WE, is formally similar to the model (5.1) broadly applied in different domains of engineering. Models (5.1) and (5.4) are in duality.

For Q = 0 the wave equation became:

$$\frac{\partial Y}{\partial T} \oplus V \otimes \frac{\partial Y}{\partial Z} = 0 \tag{5.6}$$

It is the so-called convection or plug-flow model.

A solution that can be verified directly with (5.6) is the characteristic:

$$Y = Z \oplus V \otimes T \tag{5.7}$$

Table 5.4 Convection GF(2)

T → Z ↓	0	1
0	0	1
1	1	0

Notice that V ⊕ V = 0 in GF(m). We can verify this for m = 2, 4 and 8 using Tables 5.1, 5.2 and 5.3.

For velocity V = 1 the solutions of (5.7) is Y = Z ⊕ T and has been shown in Tables 5.4, 5.5 and 5.6. These tables show convection model solutions.

Here T corresponds to steps, 0, 1, 2 and so on while Z corresponds to beds denoted 0, 1, 2 and so on and also b0, b1, b2 and so on.

The solutions of the WE with GF elements are Latin squares frequently used as DoE (Schipp et al. 1990; Iordache 2021).

The model (5.4) shows how to make use of the real model (5.1).

To face complexity we need both S1 and S2 systems since we need both cyclic and linear time for modeling (Gould 1987; Overton 1994).

PSA operation pertains to complexity domain since the number of interconnected operations and the number of possible schemes manifestly increases with the number of beds (Tondeur et al. 1985; Chiang 1988; Smith and Westerberg 1990).

Illustrative examples of scheduling will be presented in the following.

A basic flowsheet is the Two-Beds, Two-Steps system shown in Table 5.7.

The considered steps or operations are:

Table 5.5 Convection GF(4)

T → Z ↓	0	1	2	3
0	0	1	2	3
1	1	0	3	2
2	2	3	0	1
3	3	2	1	0

Table 5.6 Convection GF(8)

T → Z ↓	0	1	2	3	4	5	6	7
0	0	1	2	3	4	5	6	7
1	1	0	3	2	5	4	7	6
2	2	3	0	1	6	7	4	5
3	3	2	1	0	7	6	5	4
4	4	5	6	7	0	1	2	3
5	5	4	7	6	1	0	3	2
6	6	7	4	5	2	3	0	1
7	7	6	5	4	3	2	1	0

Table 5.7 Two-beds two-steps

b0	0	1
b1	1	0

- 0-adsorbtion
- 1-desorbtion

The beds are denoted b0 and b1.

For the Oxy-Rich processes the four steps in PSA system were:

- 0-adsorption
- 1-cocurrent production and produce purge gas for operation 3
- 2-countercurrent depressurization
- 3-countercurrent purge with gas from operation 1

The cyclic schedule proposed by Sircar and Kratz (1988) is shown in Table 5.8.

The beds are denoted b0, b1, b2 and b3.

Two beds are necessary for a feasible schedule due to the bed interconnection operations.

Operations 1 and 3 must always occur at the same time. So, the selected two beds will be b0 and b2.

The following 8 steps for PSA schedule have been considered by Smith and Westerberg (1990) and Chiang (1988).

The operations are:

- 0-adsorbtion
- 1-production
- 2-repressure operation 6 with product
- 3-pressure equalization with low-pressure operation 5
- 4-countercurrent purge with stored product
- 5-pressure equalization with high-pressure operation 3
- 6-pressure with product from operation 2
- 7-repressure with feed gas

Table 5.9 illustrates Eight-Beds Eight-Steps process

Operation 2 is coupled to operation 6, so we will select the two beds b0 and b4.

Smith and Westerberg considered that there are 7 steps distributed along 8 shifts. But the first step, adsorption and production has a double duration if compared with the duration of other steps. So, we can replace the operation 1 of Smith and Westerberg

Table 5.8 Four-beds four-steps

b0	0	1	2	3
b1	1	0	3	2
b2	2	3	0	1
b3	3	2	1	0

Table 5.9 Eight-beds eight-steps

b0	0	1	2	3	4	5	6	7
b1	1	0	3	2	5	4	7	6
b2	2	3	0	1	6	7	4	5
b3	3	2	1	0	7	6	5	4
b4	4	5	6	7	0	1	2	3
b5	5	4	7	6	1	0	3	2
b6	6	7	4	5	2	3	0	1
b7	7	6	5	4	3	2	1	0

by the operations 0 and 1 that is adsorption and production in our notation. This makes the system from Table 5.9 and the Smith and Westerberg system similar.

It is possible to run only some beds or rows from the different Tables.

We select just two beds, namely b0, b1 for GF(2), b0, b2 for GF(4) and b0, b4 for GF (8) and so on. The selection of beds is based on the imposed operations coupling.

5.2 Roadmaps

The Roadmap for PSA process design is shown in Fig. 5.1.

The strategy of doubling the number of steps in successive operation schemes offers the progression 1, 2, 4, 8 steps and so on.

The road starts from operation 0 represented as a 0D dot. Included is an experiment considered as operation 0.

The next experiment has been represented as One-Beds, b0 performing operation 0 of adsorption and then performing operation 1 of desorption or production. It is a 2 × 1 matrix corresponding to the absence of cyclicity

This shows a 2 × 2 matrix of values corresponding to two steps of the two beds b0 and b1

The 8D is partially represented to avoid complicated drawings.

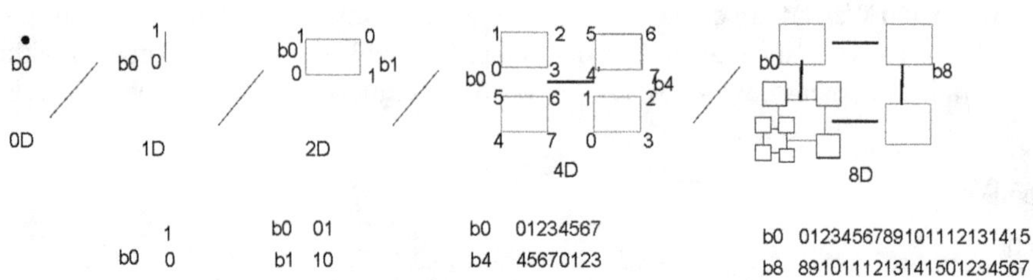

Fig. 5.1 Roadmap for higher dimensional dual process design

In the 8D frame each step of 4D is replaced by 4 component sub-steps. These sub-steps have 4 sub-sub-steps. In this way a 4D of 4D frame is resulting. The dimension is 8D = 4D + 4D.

5.3 Integration

For null velocity V the first order wave equation, WE, reduces to the kinetic model:

$$\frac{\partial Y}{\partial T} \oplus Q(Y) = 0 \tag{5.8}$$

The solution is written:

$$Y = \int Q \, dT = Q * T \tag{5.9}$$

This integral appears as a particular type of product "*".

Let us consider Y, Q, T in their binary representation (MacKay 2003; Iordache 2021). This outlines the involved scales.

We use the time $T = t_0 t_1 \ldots t_j$ interpreted as degree of classification or development, the output $Y = y_0 y_1 y_2 \ldots y_j$, the interaction rate $Q = q_0 q_1 q_2 \ldots q_j$.

Here t_j, y_j, q_j are digits, 0 or 1.

Let us define the integral "*":

$$Y = \left[\sum_j q_j t_j \right] \bmod 2 \tag{5.10}$$

An effective coupling of "1"s on the scales of Q and T is considered by this definition.

The integral definition with the mod 2 operation was selected to ensure the logical binary interpretation of the output Y (Iordache 1996). It means that two couplings cancel reciprocally to avoid inappropriate coupling of rates and time scales.

Table 5.10 shows the results of the kinetic model for 4 stages.

Table 5.10 Kinetic model (Q * T) for 4 stages

Q * T	00	01	10	11
00	0	0	0	0
01	0	1	0	1
10	0	0	1	1
11	0	1	1	0

Replacing 0 by 1 and 1 by −1 in Table 5.10 we obtain an equivalent solution in the form of Walsh–Hadamard waves. As the Latin squares these are distinctive DoE.

Table 5.11 shows the results of the modified kinetic model for 4 stages.

Table 5.12 shows the results of the kinetic model for 8 stages.

Replacing in this table, 0 by 1 and 1 by −1 we obtain the Table 5.13.

Table 5.13 shows the results of the modified kinetic model for 8 stages.

An application of Table 5.13 for processes design is presented in what follows.

The distribution of heating and cooling areas in a distillation column is the illustrating case study presented here.

We could apply the above integration method for entropy production studies.

Let y denotes the entropy and t the time.

The entropy production is governed by the equation:

$$p(y) = \frac{\partial y}{\partial t} + q(y) \geq 0 \tag{5.11}$$

Here p(y) is the entropy production q(y) is the entropy flow through the surface of the considered system (Prigogine 1955).

We use the entropy production $P = p_0 p_1 \ldots p_j$, the time $T = t_0 t_1 \ldots t_j$, the entropy $Y = y_0 y_1 y_2 \ldots y_j$, the entropy flow $Q = q_0 q_1 q_2 \ldots q_j$. Here p_j, t_j, y_j, q_j are digits, 0 or 1.

We replace the real field p(y), t, y and q by vectors P(Y), T, Y and Q and the Eq. (5.11) by (5.12):

Table 5.11 Modified kinetic model (Q * T) for 4 stages

1	1	1	1
1	−1	1	−1
1	1	−1	−1
1	−1	−1	1

Table 5.12 Kinetic model (Q * T) for 8 stages

Q * T	000	001	010	011	100	101	110	111
000	0	0	0	0	0	0	0	0
001	0	1	0	1	0	1	0	1
010	0	0	1	1	0	0	1	1
011	0	1	1	0	0	1	1	0
100	0	0	0	0	1	1	1	1
101	0	1	0	1	1	0	1	0
110	0	0	1	1	1	1	0	0
111	0	1	1	0	1	0	0	1

Table 5.13 Modified kinetic model (Q * T) for 8 stages

Q * T	000	001	010	011	100	101	110	111
000	1	1	1	1	1	1	1	1
001	1	−1	1	−1	1	−1	1	−1
010	1	1	−1	−1	1	1	−1	−1
011	1	−1	−1	1	1	−1	−1	1
100	1	1	1	1	−1	−1	−1	−1
101	1	−1	1	−1	−1	1	−1	1
110	1	1	−1	−1	−1	−1	1	1
111	1	−1	−1	1	−1	1	1	−1

$$P(Y) = \frac{\partial Y}{\partial T} + Q(Y) \geq 0 \tag{5.12}$$

Heat integration within distillation trains is a problem of engineering interest in which entropy play a significant role (Andersen 2002; Salamon et al. 2001).

How to place few thermally active stages along a distillation column is an issue that may be addressed with the help of solutions shown in Table 5.13.

Three scales for separation have to be considered in this case.

They correspond to the entire distillation column, to distances between distributed heating and cooling elements and to the trays.

According to the Table 5.13 the entropy flow Q, takes the minimum non-null value, Q = 001. This corresponds to the situation in which only one-scale, the 3rd, is involved in entropy flow process. To this solution corresponds the minimum entropy flow, and minimum entropy production MinEP.

The Table 5.14 shows an example in which there are 16 tray distillation columns.

They have been grouped in 8 pairs of trays since Table 5.13 contain only 8 columns.

The minimum entropy production, MinEP corresponds to entropy flow Q = 001.

The maximum entropy production, MaxEP corresponds to entropy flow Q = 111.

The entropy flow implies in this case all the three available scales. The heated trays are in the groups (1, 2) (5, 6) (9, 10) (13, 14) for MinEP. These groups are bolded and underlined in Table 5.14.

Table 5.14 Entropy Y(T)

Q = 001	**1**	−1	**1**	−1	**1**	−1	**1**	−1
MinEP	**1, 2**	3, 4	**5,6**	7,8	**9,10**	11,12	**13,14**	15,16
Q = 110	**1**	**1**	−1	−1	−1	−1	**1**	**1**
HighEP	**1,2**	**3,4**	5,6	7,8	9,10	11,12	**13,14**	**15,16**
Q = 111	**1**	−1	−1	**1**	−1	**1**	**1**	−1
MaxEP	**1, 2**	3, 4	5, 6	**7, 8**	9, 10	**11, 12**	**13, 14**	15, 16

This may be compared with the 1, 6, 11, 15 optimal tray locations for a 19 trays example studied by detailed models by Andersen (2002). The differences are of one tray only.

The MinEP distribution corresponds to a kind of equipartition, a situation when local entropy production is uniformly distributed over the space or time variables of the process.

Close to MaxEP is a solution that corresponds to entropy flow Q = 110 and the heated trays as: (1, 2) (3, 4) and (13, 14) (15, 16). The groups are bolded and underlined in Table 5.14. This is a situation with column extremity heated that is close to adiabatic column.

The maximum production of entropy MaxEP corresponds to Q = 111 and the heated trays: are in the groups (1, 2) (7, 8) (11, 12) (13, 14). The groups are bolded and underlined in Table 5.14.

Observe that WE offer a quick solution for heating and cooling. This is an S1 creative design. The detailed S2 analysis is based on thermodynamical data (Andersen 2002).

A distillation column may have active heating elements on each of its trays.

For this distillation column we may run DOE that is experiments from the Tables 5.13 or 5.14. Any row in such tables is an experiment that is to another variant of heating. They contain the more condensed experiments to be done to establish the desired heating element positioning.

5.4 Divergence and Convergence

Divergence and convergence is a design method that emphasizes experimentation, and iteration (Liu et al 2003). It is a human-centered design approach that focuses on understanding, needs, behaviors, and motivations to design creative solutions that meet those needs effectively.

This design process cycles between expansive and focused thinking to better understand problems people face and develop a solution. These cycles of thinking provide the structure for the iterative thinking process.

Divergence creates space and possibilities, while convergence brings focus and direction. Designers move forth and back between these two processes throughout the design thinking process, building on the insights and ideas generated during each cycle to select refine and improve the solutions.

Divergent and convergent phases are non-linear, meaning that when facilitating this process, you may move back and forth as you discover new insights during convergence.

Divergence aims exploring and generating a wide range of ideas and possibilities without excessive judgment or criticism. Divergent thinking allows designers to expand the possibilities and consider different perspectives, angles, and dimensions of the problems.

Convergent thinking refers to the process of selecting, evaluating, and refining the most promising ideas from the divergent phase. It involves narrowing down and refining possibilities by selecting the best ideas and combining them to create a coherent and effective solution.

Figure 5.2 shows Polytopic Roadmap for Divergent and Convergent design

It is considered that there are four main stages of the design (Bender and Gericke 2020):

- Tasks-T
- Conceptual design-C
- Embodiment design-E
- Detail design-D

The involved dimensions are:

- 0D-Tasks
- 1D-Concepts
- 2D-Evolvable
- 4D-Self-Evolvable
- 8D-SE of SE

For 4D frame the direct sequence T → C → E → D corresponds to Convergence while the reverse sequence D' → E' → C' → T' corresponds to Divergence.

We use the minuscule t, c, e, d to denote the control of Convergent sequence.

We use the minuscule d', e', c', t' to denote the control of Divergent sequence.

The 4D polytope shown in Fig. 5.2 illustrates the architecture of potentially self-evolvable systems. It is a representation consisting from a pair of 3D-cubes.

The second cube is a digital twin of the first cube.

The 4D-cube is obtained by joining all corners of the first 3D-cube with the corresponding corners of the second 3D-cube. The 8 lines connecting the two 3D-cubes are represented in Fig. 5.2 by a thick hyper-line. In this way the 4D-cube has a simplified representation (Thiel 1994). This kind of representation is useful for 8D polytope too.

A thick hyper-line connects the two 7D-cubes generating the 8D polytope.

Figure 5.3 shows more details and another representation of the Polytope 4D for Divergent and Convergent design

The notations associated to control that is to Self in Fig. 5.3 are:

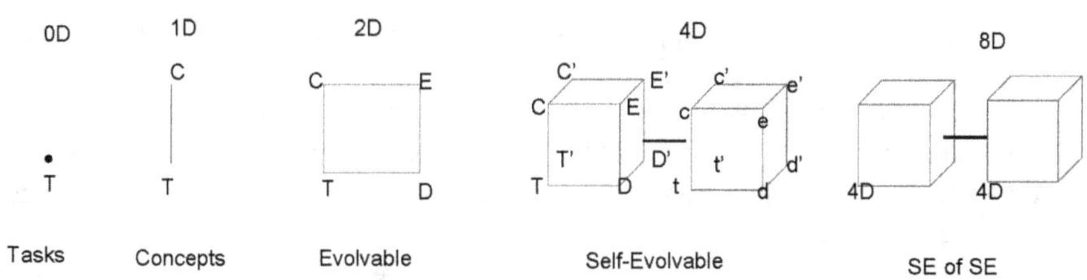

Fig. 5.2 Polytopic roadmap for divergent and convergent design

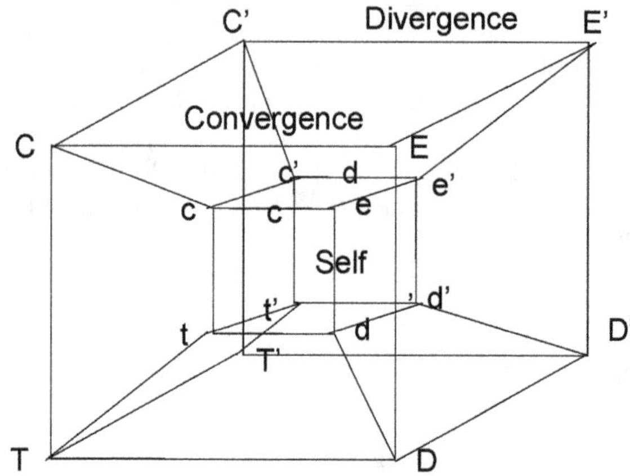

Fig. 5.3 Polytope 4D for divergent and convergent design

- t—Tasks control
- c—Conceptual Design control
- e—Embodiment Design control
- d—Detail Design control

The polytope 4D outlines that in interaction with the converging sequence T → C → E → D we need to look at the diverging sequence D′ → E′ → C′ → T′.

The inner cube shown in Fig. 5.3 marked also as Self allows dialogue and interaction of the two ways in duality relation.

The Self needs to endorse the dialogue, to ensure coordination and coherence of evolution on different levels. It results from Fig. 5.3 that the structure of the inner cube is isomorphic to that of the outer cube. Observe that the two cubes of 4D-cube are reunited in a compact way. The 4D-cube may be considered as a two scale frame.

After any step in the Convergence sequence, the systems make contact with the Self represented by the inner cube. Signals as t, c, e and d reinforce the evolution according the Convergence sequence. If the Self change the direction shifting to d', e', c', d' the next step will be along the Divergence sequence.

Polytopes as presented in Fig. 5.3 have a general applicability as reference architectures since they take into account the inherent duality of complex systems.

Observe that duality is presented for both cubes the outer and the inner one.

To make a product/system development; it is necessary to understand keys effective factors before the technical requirements can be drawn.

A proposal for a systematic approach to designing technical systems and products is presented in the following.

Figure 5.4 shows the Polytope 8D for Divergent and Convergent design

This guideline aims to construct the approach for design, which is applicable to a wide variety of tasks, and transcends specific branches of industry (Bender and Gericke 2020).

A systematic approach to the design of technical systems and products is explained in the Fig. 5.4. This figure replaces the vertices from Fig. 5.3 by 4D frames.

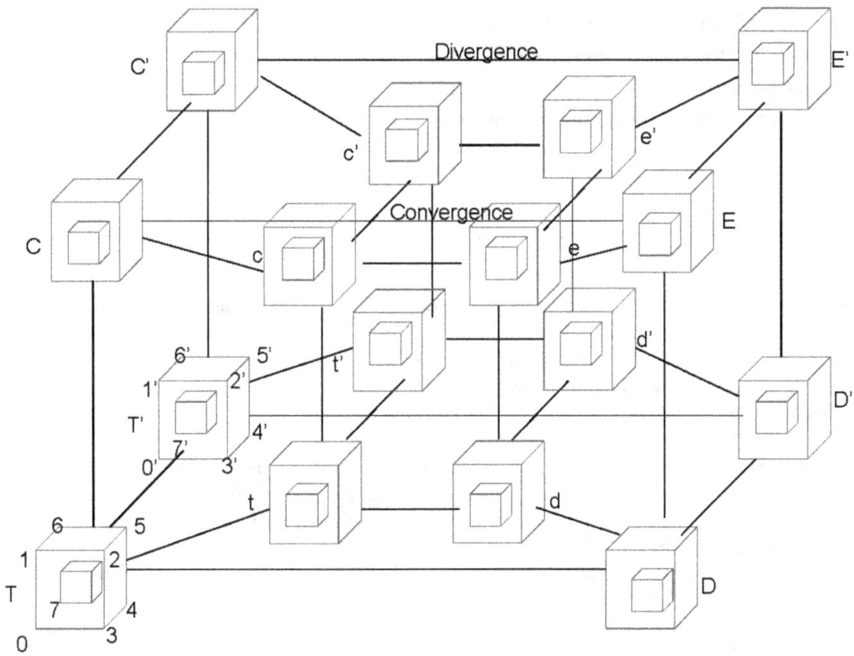

Fig. 5.4 Polytope 8D for divergent and convergent design

Above all the objective must be defined, that is, the Task.

After that the Task must be pursued in the following steps:

- 0 Clarify and define the Task (on the specification);
- 1 Determine functions and their structures (functions structure);
- 2 Search for solution principles and their combinations (principle solution);
- 3 Evaluate and select solutions concepts
- 4 Divide into realizable modules (module structure);
- 5 Develop layouts of key modules (preliminary design);
- 6 Complete overall layout (definitive design)
- 7 Prepare production and operating instructions.

The idea is to integrate of recycling-orientation of product/system during these design steps/phases. In each step when the idea/goal is not satisfy then the

The sequence 0–7 corresponds to activities module T.

The module T' includes 0'–7' corresponding to results.

Similar sequences should be proposed for other modules shown in Fig. 5.4.

One of the main aspects of this model is its intended flexibility that addresses the need to adapt high level procedural process models to the specific context of a company/design team. The model provides an overview of important factors that affect the individual adaption, resulting in context specific instances of the generic process model as shown in Fig. 5.4. Contextual factors can be distinguished into external and internal factors. External factors help to describe influences of legislation, market and customer. Internal factors describe influences of, e.g., company, product, team and the individual designer. The guideline also provides some examples of such adaptations for different contexts.

References

Andersen, T.R.: Optimal design and operation of process integrated distillation. Ph.D. thesis, Insitut for Kemiteknik, DTU, Denmark (2002)

Bender, B., Gericke, K. (eds.): Pahl/Beitz Konstruktionslehre, 9th edn. Springer, Berlin (2020)

Cash, P., Daalhuizen, J., Valgeirsdottir, D., Van Oorschot, R.: A theory-driven design research agenda: exploring dual-process theory. In: Proceedings of the 22nd International Conference on Engineering Design (ICED19), Delft, The Netherlands, pp. 5–8 (2019)

Chiang, A.S.T.: Arithmetic of PSA process scheduling. A.I.Ch.E.J. **34**(11), 1910–1912 (1988)

Croskerry, P.: A universal model of diagnostic reasoning. Acad. Med. **84**(8), 1022–1028 (2009)

Evans, J.St.B., Stanovich, K.E.: Dual process theories of higher cognition: advancing the debate. Perspect. Psychol. Sci. **8**, 223–241 (2013)

Gould, S.J.: Time's Arrow, Time's Cycle: Myth and Metaphor in the Discovery of Geological Time. Harvard University Press, Cambridge (1987)

Grothendieck, A.: Pursuing stacks (1983)

Houdé, O.: 3-System Theory of the Cognitive Brain, A Post-Piagetian Approach to Cognitive Development. Routledge, London (2019)

Iordache, O.: Dyadic frames for intermittency. Perturbed models. In: Bayod, J.M., De Grande De Kimpe, N., Schikhof, W.H. (eds.) Proceedings of the Conference on p-adic Functional Analysis, Laredo, (Spain) May 1990. Lecture Notes in Pure and Applied Mathematics, vol. 137, pp. 89–99, Marcel Dekker, New York (1992)

Iordache, O.: Theoretical frames for smart structures. Mater. Sci. Eng. C **4**(3), 143–148 (1996)

Iordache, O.: Polystochastic Models for Complexity. Springer, Berlin (2010)

Iordache, O.: Polytopic Roadmaps. Springer, Cham (2021)

Iordache, O., Scarlat, R., Frangopol, P.T.: Intermittent oscillations in model biomembranes. Rev. Roum. Biochim.roum. Biochim. **27**, 43–49 (1990)

Iordache, O., Valentin, G., Corriou, J.P., Pons, M.N., Pethö, A.: Intermittent interfacial transfer. a dyadic model. Acta Chim. Hungar. Models Chem. **1**(130), 1–18 (1993)

Kahneman, D.: A perspective on judgement and choice. Am. Psychol. **58**(9), 697–720 (2003)

Kahneman, D.: Thinking, Fast and Slow. Farrar, Straus and Giroux, New York (2011)

Lidl, R., Niederreiter, H.: Finite Fields, 2nd edn. Cambridge University Press (1997)

Liu, Y.C., Chakrabarti, A., Bligh, T.: Towards an 'ideal' approach for concept generation. Des. Stud. **24**, 341–355 (2003)

Lupasco, S.: Le principe d'antagonisme et la logique de l'énergie. Prolégomènes à une science de la contradiction. Hermann & Co., Paris (1951)

MacKay, D.: Information Theory, Inference and Learning Algorithms. Cambridge University Press (2003)

Moore, D., Sauder, J., Jin Y.: A dual-process analysis of design idea generation. In: Proceedings of the ASME 2014 International Design Engineering Technical Conferences & Computers and Information in Engineering Conference IDETC/CIE 2014. Buffalo, New York, USA (2014)

Overton, W.F.: The arrow of time and the cycle of time: concepts of change. Cogn. Embodiment Psychol. Inquiry **5**(3), 215–237 (1994)

Overton, W.F., Ricco, R.B.: Dual-systems and the development of reasoning: competence-procedural systems. In: Wiley Interdisciplinary Reviews: Cognitive Science (2010)

Prigogine, I.: Introduction to Thermodynamics of Irreversible Processes. Interscience, NY, USA (1955)

Rhee, H.K., Aris, R., Amundson, N.R.: First Order Partial Differential Equations II, Theory and Applications of Hyperbolic Systems of Quasilinear Equations. Prentice-Hall, Englewood Cliffs, NJ (1989)

Ruthven, D.M.: Principles of Adsorption and Adsorption Processes. Wiley, New York (1984)

Ruthven, D.M., Farooq, S., Knaebel, K.S.: Pressure Swing Adsorption. VCH, New York (1994)

Salamon, P., Nulton, J.D., Siragusa, G., Andersen, T.R., Limon, A.: Principles of control thermodynamics. Energy **26**, 307–319 (2001)

Schipp, F., Wade, W.R., Simon, P.: Walsh Series. An Introduction to Dyadic Harmonic Analysis. Adam Hilger, New York (1990)

Sircar, S., Kratz, W.C.: A pressure swing adsorption process for production of 23–50% oxygen-enriched air. Sep. Sci. Technol. **23**(4–5), 437–450 (1988)

Smith, O.J., Westerberg, A.W.: Mixed-integer programming for pressure swing adsorption cycle scheduling. Chem. Eng. Sci. **45**(9), 2833–2842 (1990)

Thiel, T.: The design of the connection machine. Des. Issues **10**, 1, 5–18 (1994)

Tondeur, D., Migault, G., Wankat, P.C.: Gas purification by pressure-swing adsorption. Process Anal. Entropie **123**, 39–55 (1985)

Chapter 6
Industry

Abstract Roadmaps and frameworks for Industry 4.0 and Industry 8.0 have been introduced here. The Industry 4.0 is an attempt of unification between physical assets and advanced digital technologies. A generic procedure to introduce Industry 4.0 in Small and Medium Enterprises, SME, was introduced. The 4D frameworks are associated to Industry 4.0. The Reference Architectural Model RAMI 4.0 give companies a framework for developing future products and business models. It is a map showing how to approach the deployment of Industry 4.0 in a structured manner. Industry 8.0 and RAMI 8.0 as described here are 8D frameworks. The Polytopic Roadmaps are correlated to dimensionality of reference architectures. Chemical engineering paradigms are significant case studies of the 8D Program.

6.1 Industry 4.0

Industry 4.0 is a vision of integrated industry implemented by computing, software, and internet technologies. The label 4.0 refers to the vision of the fourth industrial revolution.

Industry 4.0 is increasingly entering the industrial landscape of Small and Medium Enterprises, SME. A generic procedure to introduce Industry 4.0 in SME was developed. The purpose of this procedure is to give SME guidance on how to address Industry 4.0 in order to identify enterprise-specific technology solutions, to improve existing processes, to advance existing business models and to exploit new business models (Anderl et al. 2015). The development of toolboxes for engineering follows the Guideline Industry 4.0 set by the VDMA (Verband Deutscher Maschinen- und Anlagenbau).

To meet this generic method, applications are firstly defined and divided in development stages. Relevant applications are identified based on the defined engineering tasks and potentials areas. Furthermore, the applications are classified respecting the categories of engineering areas.

The most important applications have to be selected and used in toolboxes.

O. Iordache, *General Reference Architecture Frameworks*, Studies in Systems, Decision and Control 73, https://doi.org/10.1007/978-3-031-70718-6_6

Toolbox tables are structured in applications vertically (y-axis) and development stages horizontally (x-axis). Application displays Industry 4.0 themes. Every application is broken down in five technological and sequential stages.

The lowest level denoted here 0 corresponds to the conventional non-industrial stage. The highest level denoted here 4 represent the Industry 4.0 perspective.

Several applications may be grouped in areas or domains. Examples of areas of applications are Product, Production, Business Models, Intra-logistics, Assembly and Security, Engineering and so on. We could range these domains in the natural order for product implementation starting from design to testing.

Table 6.1 shows an example of Industry 4.0 toolbox.

Application areas describe topics of interest in the context of Industry 4.0, for industrial product development. The four rows in Table 6.1 refer to Product.

Performance levels describe technological approaches which are structured into a sequence reaching from none or low level industry performance labeled 0 to high level Industry 4.0 performance labeled 4.

The abbreviated notations for applications are:

- Integration-I
- Communication-C
- Monitoring-M
- IT Service-IT

The development stages have been identified from 0 to 4.

The first column from Table 6.1 corresponds to the specific applications.

Consider for instance that the application is the integration of sensors and actuators.

The development stage starts with the product having no use of sensors and actuators -stage 0. Industry 4.0 foresights are that the product responds independently based on gained data-stage 4. The intermediary stage 1-sensors integrated stage, stage 2-sensor reading and stage 3-data evaluated for this application are established on the

Table 6.1 Industry 4.0 toolbox. Product

Application	0	1	2	3	4
Integrationsensors/ actuators **I**	No sensors/ actuators	Sensors/ Actuators integrated	Sensors reading processed by system	Data evaluated analyzed by system	System independently respond based on data analysis
Communication connectivity **C**	No interface	I/O signals	Field bus interface	Ethernet interface	Product access to internet
Monitoring **M**	No monitoring	Failure detection	Recording operating conditions	Prognosis of condition	Independent control measures
Product related IT service **IT**	No services	On line portals	Service direct via product	Independent performed services	Complete integration in infrastructure of IT services

basis of the expertise in the domain. Practical expertise allows defining the qualitative progress steps for the specific domain of application. These steps may be correlated to singles, mappings, systems, SoS and so on in development studies (Fischer 1980). This highlights the parallelism between virtuality and reality a repeated finding of this book.

The next application in Table 6.1 is the communication and connectivity.

The development starts with the product having no interfaces for communication-stage 0. Industry 4.0 vision is that of the product having access to the internet-stage 4.

The next example of application is the monitoring. The development starts with no monitoring for the product-stage 0. Industry 4.0 vision corresponds to the independently adopted control measures of the product-stage 4.

The last example of application is the product-related IT Services. The development stage starts with no services. Industry 4.0 vision is that of the complete integration into an infrastructure of IT Services.

Figure 6.1 shows the polytope 4D for Industry 4.0 Product.

A 4D polytope may be associated to any specific application, I, C, M or IT.

The 4D polytope outlines two sequences for evolution, the Direct sequence $0 \rightarrow 1 \rightarrow 2 \rightarrow 3 \rightarrow 4$ that corresponds to the upgrading stages and the Reverse sequence $4' \rightarrow 3' \rightarrow 2' \rightarrow 1' \rightarrow 0'$ that corresponds to the downgrading modified stages.

This polytope is useful for grafting strategy of development.

The Direct acquired development is questioned and modified by the Reverse sequence.

This corresponds to an Iteration step.

The stages 0, 1, 2, 3, 4 may be associated to upgrading.

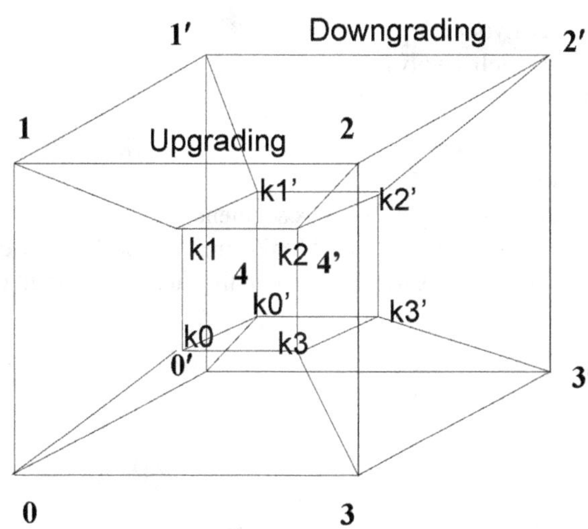

Fig. 6.1 Polytope 4D for industry 4.0 product

The stages 0', 1', 2', 3', and 4' correspond to downgrading and may be associated to dual or modified stages. The modifications means that once the application arrived at stage 4 all the preliminary stages will be modified (digitalized in some cases).

The Self and Self' that is 4 and 4' are naturally associated to the Included Middle.

Self is represented as two squares, the front face or 4 and the back face or 4', with corners connected. This allows integrating Direct and Reverse sequences. Details about 4 and 4' are required to complete the polytope 4D shown in Fig. 6.1. For instance, we consider for stage 4 the sub-levels: k0, k1, k2 and k3 and for 4' the sub-levels: k0', k1', k2' and k3'.

Observe that the Polytope 4D as shown in Fig. 6.1 extends the information from Table 6.1.After the stage, 0, 1, 2, 3 and 4, the dual or modified stages 0', 1', 2', 3' and 4' are included. Moreover a split of 4 and 4' in control stages is assumed.

6.2 Roadmaps to 8.0

A case study of leap forward from Industry 4.0 to Industry 8.0 is presented here.

We consider an example of industry 4.0 implementation that needs the Direct sequence of steps: Integration → Communication → Monitoring → IT Service.

This is suggested by Table 6.1.

The Direct sequence is: I → C → M → IT.

Figure 6.2 shows the Polytopic Roadmap for Industry 8.0 Product

The dimensions illustrated in Fig. 6.2 are:

- 0D-Integration
- 1D-Communication
- 2D-Evolvable
- 4D- Self-Evolvable
- 8D-SE of SE

The 0D is associated to Integration, I. The 1D includes I and also the Communication, C.

The 1D is represented as segments, IC.

Putting together I, C, Monitoring M, and IT Service IT, a 2D evolvable system may result. Apostrophes 'denote the dual or modified components, IT', M', C', I'.

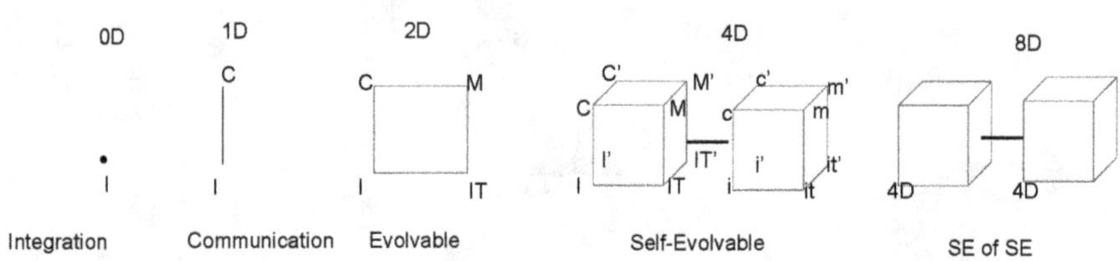

Fig. 6.2 Polytopic roadmap for industry 8.0 product

The sequence I → C → M → IT is modified to the Reverse sequence IT' → M' → C' → I'.

To accomplish self-evolvability the system needs control. The notations associated to control in 4D in Fig. 6.2 use minuscule.

We denote:

- i—Integration control
- c—Communication control
- m—Monitoring control
- it—IT Service control

The 4D frame shows the first cube of real stages, and the second cube of control stages.

The second cube is the digital twin of the first cube. Correlating the two cubes as shown by a thick hyper-line allows self-evolvability, SE. The use of thick lines to represent 4D and 8D frames is discussed by Thiel (1994)

The term Industry 8.0 is associated to 8D frameworks. The 8D is achieved as 4D of 4D.

The 8D frame results by replacing any application in the 4D frame by the detailed 4D frame as shown in Fig. 6.2.The new frame has the dimension 8D = 4D + 4D.

So as 4D was associated to Industry 4.0 Product, the resulting 8D will be associated to Industry 8.0 Product.

Figure 6.3 shows more details and another presentation for the polytope 4D presented in the Fig. 6.2.stage 4D.

The polytope 4D outlines that in interaction with the converging, constructing or Direct sequence I → C → M → IT we need to look at the diverging, deconstructing or Reverse sequence IT′ → M′ → C′ → I′.

Since dual complementary descriptions are irreducible and in some cases contradictory, the concept integration or coordination of complementary descriptions

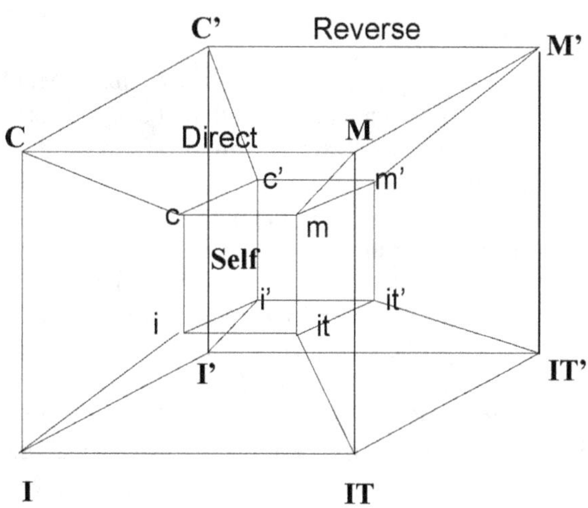

Fig. 6.3 Polytope 4D for industry 8.0 product

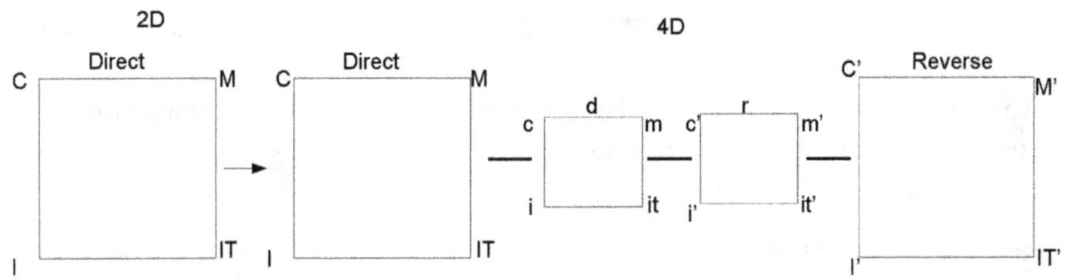

Fig. 6.4 Developed roadmap for polytope 4D for industry 8.0 product

requires a higher level description, a new dimension that emerges as a new hierarchical level of complexity. There are the cases of 4D and 8D reference architectures.

The inner cube, marked as the Self allows coherent dialogue and interaction of the two ways in duality relation. These ways correspond to construction or Direct way, from I to IT, or to or Reverse way from IT′ to I′.

The Self needs to endorse the dialogue, to ensure coordination of evolutions on different levels. It results from Fig. 6.3 that the structure of the inner cube is isomorphic to that of the outer cube. Observe that the two cubes of 4D-cube are reunited in a compact way.

The inner cube is a digital twin of the outer cube. The 4D-cube may be considered as a two scale frame.

After any step in the Direct sequence, the systems make contact with the Self represented by the inner cube. Signals as i, c, m and it, reinforce the evolution according the Direct sequence. If the Self have to change the direction shifting to it', m', c', i' the next step will be along the Reverse sequence. The set (i', c', m', it') corresponds to the control of the Reverse sequence.

Observe that duality is presented for both cubes the outer and the inner one.

Figure 6.4 shows a developed roadmap to polytope 4D for Industry 8.0 Product.

The 2D and 4D frames are presented in a developed form here. Direct impose Reverse sequence.

In this step a new frame, the Self or Included Middle is considered.

The Self is shown with Direct d, part and Reverse r, part.

Since Direct and Reverse frames are 2D the d and r frames are 2D too. They illustrate the conditions.

Relating d and r allows obtaining the inner cube. The 4D frame, by inclusion the inner cubic frame in the outer cubic frame. It will be a 4D-cube representing the Self-Evolvable systems. The inner cube is a digital twin of the outer cube.

The developed roadmap may be interpreted in terms of dual process theory (Kahneman 2011). The Direct way $I \rightarrow C \rightarrow M \rightarrow IT$ pertains to System 1 processes while the Reverse way $IT' \rightarrow M' \rightarrow C' \rightarrow I'$ pertains to System 2 processes.

Lupasco (1951), Houdé (2019) go beyond the (System 1, S1) and (System 2, S2) approaches to put forward a theory in which the included third T, that is the inhibitory-control system (System 3, S3) takes precedence.

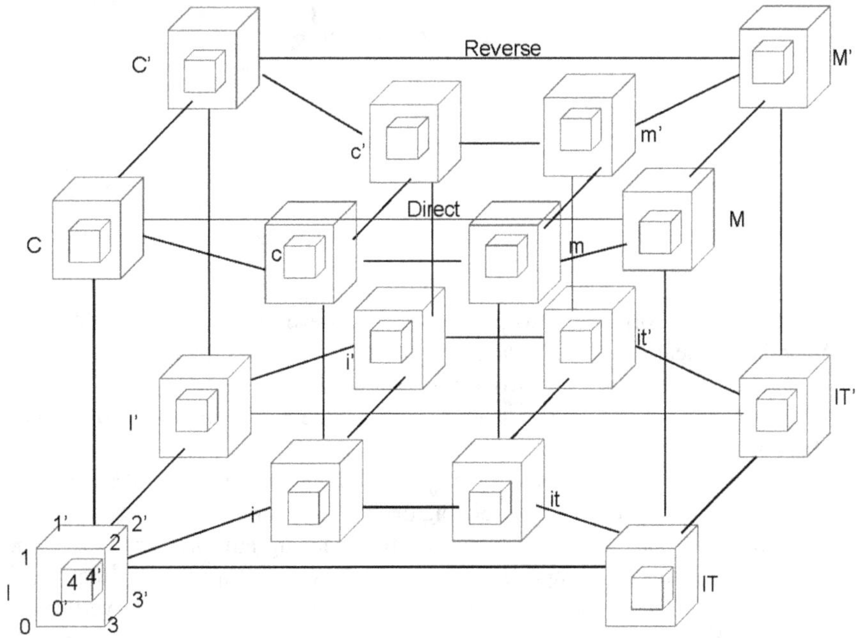

Fig. 6.5 Polytope 8D for industry 8.0 product

T or S3 denotes the bridge between S1 and S2. Here S3 is associated to the Self putting together the d and r parts of IM. S3 is represented as a 3D-Cube.

Figure 6.5 shows the polytope 8D for Industry 8.0 Product.

This is the GRAF studied in 8D Program.

A grafting strategy that maintains the 4D architecture but replace any 0D vertices by developed and structured 4D vertices was considered. It is a hypercube of hypercubes.

The new dimension is 8D = 4D + 4D. Here, the first 4D come from the initial 4D frame as shown in Fig. 6.3 and the next 4D from the component 4D frames replacing the vertices of the initial hypercube. Hypercubes as I, refer to real stages while hypercubes as i, refers to virtual control.

The presented Roadmaps outlined two sequences for evolution, the direct sequence I $\rightarrow$ C $\rightarrow$ M $\rightarrow$ IT that corresponds to the Direct stages and the reverse sequence IT$'$ $\rightarrow$ M$'$ $\rightarrow$ C$'$ $\rightarrow$ I$'$ that corresponds to the Reverse stages.

This is a sustainable type of development. The directly acquired development is questioned and modified by the reverse sequence. The bottom-up construction from direct sequence complements the top-down deconstruction of the reverse sequence.

The 4D-frames I, C, M, IT are based on the step described in Table 6.1 and Fig. 6.1.

For the module I we have to consider sub-levels as 0, 1, 2 and so on.

Modified values are denoted by 0', 1', 2' and so on (see Fig. 6.5).

For a complete description of I the elements of sub-self should be identified.

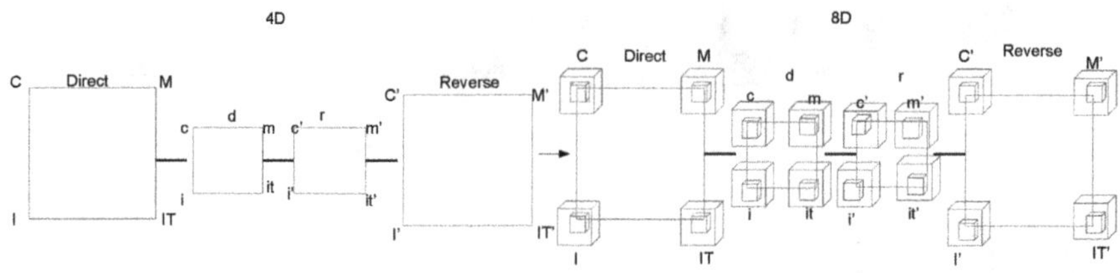

Fig. 6.6 Developed roadmap for polytope 8D for industry 8.0 product

Observe that the Polytope 8D as shown in Fig. 6.5 develops the info from Table 6.1.

After the applications, I, C, M, and IT, the dual or modified stages I', C', M', and IT' are included. Moreover a control twin is included.

Figure 6.6 shows a developed roadmap for polytope 8D for Industry 8.0 Product.

The 4D and 8D frames are presented here.

The 4D architecture shows the developed frame of an external and an internal 3D-cube. These cubes result by correlating the corresponding 2D frames.

The transition from 4 to 8D is achieved by replacing any 0D vertex from the initial 4D architecture by component 4D frames. It is a grafting strategy that keeps the initial 4D architecture but replace any 0D vertices by component 4D vertices.

It results a hypercube of hypercubes that is a 4D of 4D. The new dimension is 8D = 4D + 4D. The first 4D come from the initial 4D frame and the next 4D from the component 4D frames replacing the 0D dots of the initial 4D hypercube

In the next step the inclusion of a new frame the Self is considered. The Self is shown with Direct d, part and Reverse r, part.

Relating d and r allows obtaining the inner 7D-cube. Finally the 8D frame results, by inclusion of the inner 7D-cube in the external 7D-cube.

6.3 RAMI

The Industrial Internet is an internet of things, machines, computers and people enabling intelligent industrial operations using advanced data analytics for transformational business outcomes. The Industrial Internet Consortium (IIC) is devoted to make the Industrial Internet a reality.

The reference architecture addresses the Industrial Internet problems by providing common and consistent definitions in the system of interest, decompositions, and design patterns, and a common terminology to discuss the specification of implementations so that options may be compared (Iordache 2019).

Industry 4.0 and related initiatives recognize that efficiently building self-managing production processes requires open software and communications standards that allow sensors, controllers, people, machines, equipment, logistics systems, and products to communicate and cooperate with each other directly. Future automation systems must adopt open source interoperability software application and communication standards similar to those that exist for computers, the internet, and smart phones.

Industry 4.0 demonstrations acknowledge this by leveraging existing standards.

The harmonization of standards worldwide took another step forward when providers of the Platform Industry 4.0 and the Industrial Internet Consortium (IIC) explored the potential alignment of their two architecture efforts-respectively, the Reference Architecture Model for Industry 4.0 (RAMI4.0) and the Industrial Internet Reference Architecture (IIRA).

Significant cooperative efforts recognize that manufacturing has worldwide interdependencies requiring common standards and interoperability (OMG 2008).

Reference architectures have been presented by Uslar and coworkers (Uslar and Trefke 2014; Uslar and Engel 2015; Uslar and Gottschalk 2015).

The Reference Architectural Model for Industry 4.0, RAMI 4.0, consists of a three-dimensional coordinate system that describes all crucial aspects of industry 4.0.

In this way, complex interrelations can be broken down into smaller and simpler clusters.

The three axes are: Hierarchy Levels, HL, Life Cycle and Value Streams, LC, and Layers, L.

The Hierarchy Levels axis is based on the hierarchical levels from the international standards series for enterprise IT and control systems.

These Hierarchy Levels represent the different functionalities within factories or facilities. In order to represent the industry 4.0 environment, these functionalities have been expanded to include workpieces, labeled Product, and the connection to the Internet of Things and Services, labeled Connected World.

The Life Cycle and Value Stream axis represents the life cycle of facilities and products, based on standard for life-cycle management. Furthermore, a distinction is made between types and instances. A type becomes an instance when design and prototyping have been completed and the actual product is being manufactured.

RAMI 4.0 combine the basic elements of industry 4.0 in a three-dimensional layer model. Based on this framework, different industry 4.0 technologies can be classified and further developed.

The Layers axis show six layers on the vertical axis and serve to describe the decomposition of a machine into its properties structured layer by layer, that is, the virtual mapping of a machine. Such representations originate from information and communication technology, where properties of complex systems are commonly broken down into layers.

Within the three axes, all crucial aspects of industry 4.0 can be mapped, allowing objects such as machines, technologies, test-beds to be classified according to the model.

Highly flexible industry 4.0 concepts can thus be described and implemented using RAMI 4.0. The reference architectural model allows for step-by step migration from the present industrial stage into the world of industry 4.0.

RAMI 4.0 integrate different user perspectives and provide a common understanding of industry 4.0 technologies. With RAMI 4.0, requirements of sectors from manufacturing automation and mechanical engineering or chemical process engineering can be addressed in standardization committees. Thus, RAMI 4.0 provide the common understanding for standards and use cases.

RAMI 4.0 can be regarded also as a 3D map of industry 4.0 solutions. It provides an orientation for plotting the requirements of sectors together with national and international standards in order to define and further develop industry 4.0.

The 8D reference architecture is proposed here. It supplements the 3 axis of conventional RAMI by a supplementary axis of Assets.

Thus the basic levels or modules are:

- Assets-A
- Hierarchy Levels-HL
- Life Cycle and Value Streams-LC
- Layers-L

Figure 6.7 shows the polytope 8D for RAMI 8.0.

The main levels Assets, Hierarchy Levels, Life Cycle and Layers are modified to the dual main levels Assets', Hierarchy Levels', Life Cycle' and Layers'.

Apostrophes 'mark the dual or modified states.

Examples of details for main levels will be presented in the following.

While the Asset view shows resources available to the enterprise for instance.

- 0-Personnel

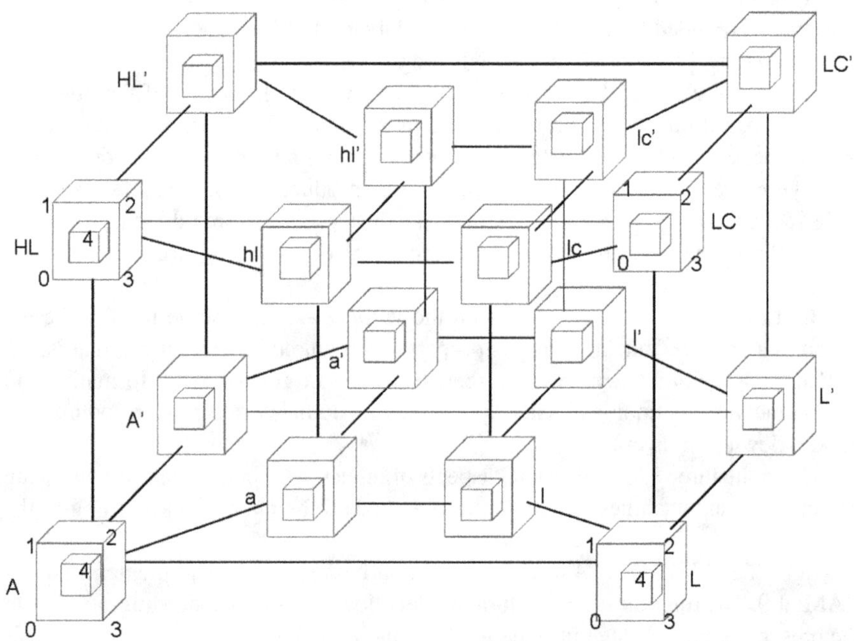

Fig. 6.7 Polytope 8D for RAMI 8.0

- 1-Processes
- 2-Products
- 3-Plants
- 4-Enterprise

Enterprise is identified as the Self of this A polytope.
The Hierarchy Levels sub-levels are.

- 0-Product Field
- 1-Control Station
- 2-Work Unit
- 3-Enterpise
- 4-Connected World

Connected world is identified as the Self of this HL polytope.
The sub-levels of Life Cycles are

- 0-Development Type
- 1-Maintenance Type
- 2-Production Instance
- 3-Maintenance Instance

Currently we don't have the identification for the Self of LC polytope
The sub-levels of the polytopic architecture for Layers are identified as follows:

- 0-Asset Integration
- 1-Communication
- 2-Information
- 3-Function
- 4-Business

Business is identified as the Self of this L polytope.

6.4 Chemical Engineering Paradigms

In the first half of the 20th century faculties of applied chemistry had to implement courses on chemical and thermo-chemical calculus. At the same time appeared the books related to chemical process principles. Such books introduced industrial chemistry as a preliminary step, a pre-paradigm for chemical engineering (Hougen and Watson 1936).

The focus was on the sciences rather than on engineering.

Instead to describe each technological process some chemists regards an industrial chemical process to be composed by distinct sections which are present in many other processes. This concept and its applications have been considered to be the first paradigm of chemical engineering, namely the unit operations (Walker et al. 1923).

The thousands of industrial chemical processes can't be individually treated to the detailed scale as imposed by design and operation of the plants. Instead, the processes are made from a much smaller number of unit operations.

The transport phenomena paradigm, an upper systematization and synthesis evolution was proposed as a second paradigm (Bird et al. 1960). The paradigm of transport phenomena approaches the three elementary physical processes, which take place in any kind of unit operation: momentum, energy, and mass transport. Unit operations can be considered as specific applications of these three fundamental transport processes. Combinations of transport processes give unit operations and combinations of unit operations give technologies.

The transport phenomena paradigm extends the content of chemical engineering to a fundamental, theoretical science, closely linked with physics, mechanics, thermodynamics and mathematics. The origin of the second paradigm was the consequence of the request for a deep, scientific knowledge of the phenomena which explain what happened inside of unit operations.

In the second half of 20th century the diversity of industrial products has a growth, and correspondingly, strong market fights have evolved between producers companies.

The same things happened with chemical products. The importance of properties and qualities of chemical products become essentially. It was observed that product engineering problems can't be solved by conventional chemical engineering approaches and paradigms. This explains why product engineering was recognized by some engineers as a third paradigm of chemical engineering. The third paradigm was imposed by the need for technical and economical products performances generated by a strong competitive market environment.

Product engineering techniques is largely based on heuristics since data are limited, followed by detailed calculations when data become available, this being the essence of the third paradigm (Hill 2009). Some basics of the third paradigm were presented by Cussler and Moggridge (2001). They proposed a generic framework for chemical product design, based on a four-step algorithm. The steps are:

(1) identify customer needs
(2) generate ideas to meet those needs
(3) select among the ideas
(4) manufacture the product.

The step four of chemical product design contains all the following four steps of chemical process design:

(1) batch versus continuous process
(2) inputs and outputs
(3) reactor and recycles
(4) separations and process integration

A paradigm considered by some engineers as the fourth one is sustainability (Clift 2006). The concept of sustainability is imposed in al human activities fields, especially in industrial domains. Chemical industry, as a strong materials and energy consumer, with a big ecological impact, could not remain outside of sustainability requirements (Stock and Selinger 2016).

The sustainability was studied by Ehrenfeld (2008) who defines it as the possibility that different life and economic forms will flourish forever. Flourishing is understood as an emergent quality of the global socioeconomic and natural systems.

The sustainable chemical engineering paradigm is set on the recognition of limitation of resources, the requirement for inter and trans-generational equity within human society and the need for preservation of life supporting natural systems (Narodoslawsky 2013).

Sustainability is defined as the development that meets the needs of the present without compromising the ability of the future generations to meet their own needs. Sustainability requires human society to live within the limitation of our planet in a way that allows infinite development in temporal terms (Heinzle et al. 2006).

Related to the previously accepted paradigms, the multiscale methodology is progressively gaining recognition and application as a new paradigm. Since long, chemical engineers recognize that chemistry on the molecular scale plays the essential role in a commercial chemical reactor, and try to incorporate it into the chemical product manufacture on the macro-scale. However, they were troubled with difficulties in developing multiscale models that they had the ability to solve and get mathematical solutions. With the development of scientific theories and instruments, many chemical engineering researchers addressed an integrated multidisciplinary and multiscale approach for chemical product engineering and process systems engineering.

Li and Kwauk (1994, 2003) used a multiscale methodology to model the circulating fluidized bed riser with incorporation of mechanisms and phenomena on two different scales and termed their energy minimization multi-scale model as a variational multiscale methodology.

Villermaux (1966), Lerou and Ng (1996) and Charpentier (2009, 2010) advocated the multiscale methodology for chemical product engineering and process system engineering as a new paradigm of chemical engineering in view of its central role for process intensification and product design and engineering. They emphasized on the computer-aided multiscale modeling and simulation of real-life projects from the molecular-scale to the production-scale. We should refer to multiscale for the real and for the virtual domains.

During the evolution of chemical engineering every new paradigm was a step forward which has extended the manifold of the tasks that can be solved. However, the older paradigms are not neglected but included. Some paradigms must be used together in order to solve the complex chemical engineering problems.

The first paradigm of unit operations and the second paradigm of transport phenomena are recognized by numerous chemical engineers. These paradigms can't be separated.

What the next paradigm is remains still an open question. Several proposals such as Chemical product engineering,Sustainable chemical engineering, Multi-scale engineering, Computational chemical engineering, Intelligent chemical engineering, have been considered as potential candidates for new paradigms.

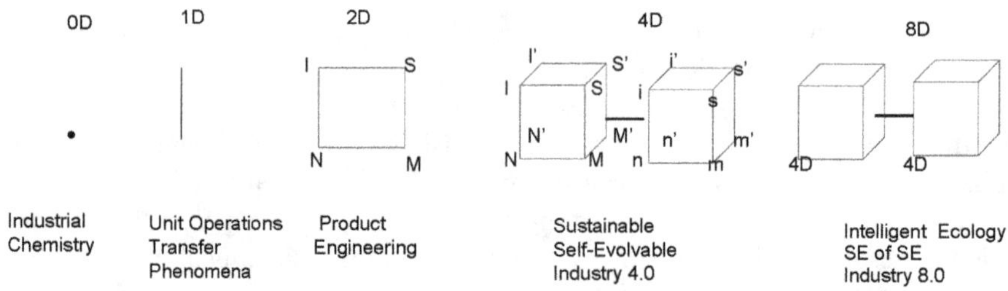

Fig. 6.8 Polytopic roadmap for chemical engineering paradigms

The Polytopic Roadmap for chemical engineering paradigm will be examined here.

It is based on the dimensionality of the reference architecture frameworks associated to conventional paradigms. The existing paradigms are redistributed under the dimensionality label, 0D, 1D, 2D, 4D and 8D, and in some cases renamed according to dimensionality of reference architectures.

Figure 6.8 shows the Polytopic Roadmap for Chemical Engineering Paradigms

The dimensions illustrated in Fig. 6.8 are:

- 0D-Industrial Chemistry
- 1D-Unit Operations. Transfer Phenomena
- 2D-Product Engineering
- 4D-Sustainable. Self-Evolvable, Industry 4.0
- 8D-Intelligent. Ecology. SE of SE, Industry 8.0

Industrial Chemistry is considered as a 0D pre-paradigm.

Unit Operations born from Industrial Chemistry are represented as 1D segment.

Transfer Phenomena are included here since combinations of transport processes give unit operations.

Several operations allow the Product Engineering and are represented as 2D frames.

As 2D frame we consider here the proposed generic framework for chemical Product design, based on a four-step algorithm (Cussler and Moggridge 2001).

The notations for Fig. 6.8 are:

- N-needs
- I-ideas
- S-selection among ideas
- M-manufacture

The apostrophes ‘denote dual or modified stages and minuscule denotes control, as follows:

- N’-dual needs
- I’-dual ideas
- S’-dual selection among ideas

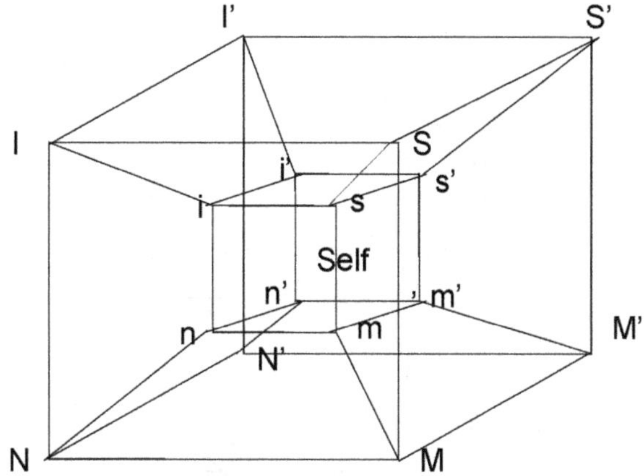

Fig. 6.9 Polytope 4D for sustainable chemical engineering

- M'-dual manufacture
- n-needs control
- i-ideas control
- s-selection among ideas control
- m-manufacture control

The 2D Product shows the direct sequence N → I → S → M.

The 4D will include Product in duality. It shows the dual 2D frame, the reverse sequence: M' → S' → I' → N'.

The second cube shown for 4D is the digital twin of the first cube shown in 4D.

It represents the control and allows sustainability.

The step M of Product contains the sub-steps of chemical process design. It is a multiscale situation.

Figure 6.9 shows details and another presentation of the Polytope 4D for Sustainable Engineering

The inner cube shown in Fig. 6.9 is the digital twin of the outer cube.

The framework 8D is resulting as 4D of 4D architectures. It results by grafting new 4D frames in any 0D vertex of the previously obtained 4D frame.

Figure 6.10 shows the Polytope 8D for Intelligent Engineering Paradigm

The step four, M, of Product design contains the four steps of chemical process design (Cussler and Moggridge 2001). It is a multiscale situation denoted by 0, 1, 2, 3 for direct subsequence and 0',1'0.2'0.3' for reverse or dual subsequence.

The notations for Fig. 6.10 are:

- 0-batch versus 0'-continuous process,
- 1-inputs versus 1'-outputs
- 2-reactor versus 2'-recycles
- 3-separations versus 3'-process integration

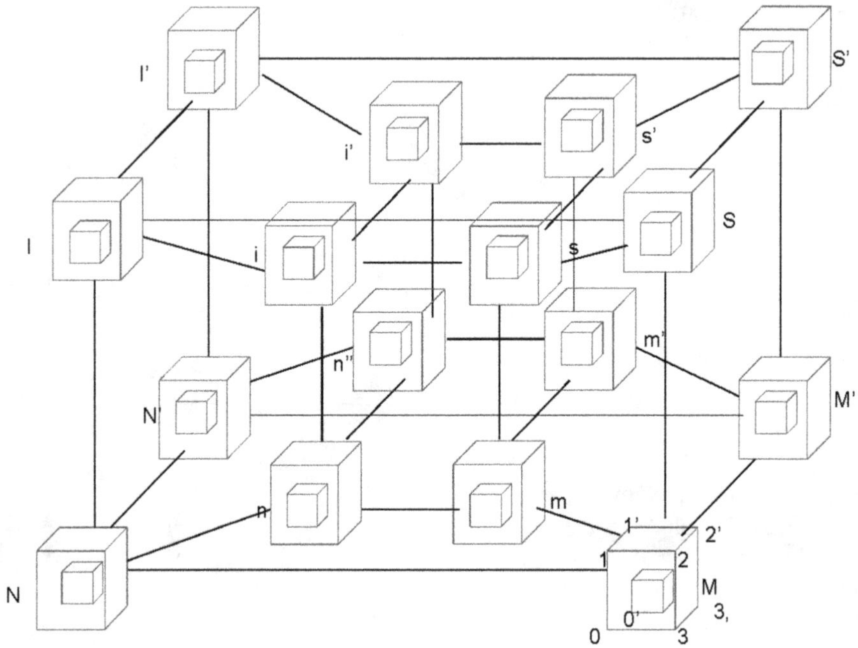

Fig. 6.10 Polytope 8D for intelligent chemical engineering

For a complete description of M the components of the sub-self of M should be identified. It refers to the control of M. Also, for 8D we need to graft 4D frames not only for module M but also for the other modules N, I and S.

The 8D frame corresponds to Intelligent Engineering paradigm.

It is related to Industry 8.0 paradigm and to Ecological paradigms.

Stephanopoulos (1986) anticipated that, concepts from artificial intelligence will play a major role in chemical engineering

Han et al. (1995) shows how one can use ideas and techniques from artificial intelligence, such as symbolic modeling, knowledge-based systems, and logic, to construct a computer-implemented model of the design process.

Based on the Douglas (1988) Conceptual Design model, Han et al. (1995) shows how to generate models of the structure of design tasks, design decisions, and the state of design, thus leading to automation of large segments of the synthesis of chemical processing schemes.

The result is a human-aided, machine-based design paradigm, with the computer knowing how the design is done, what the scope of design is, and how to provide explanations and the rationale for the design decisions and the resulting final design.

The human-aided, machine-based design may be associated to the Intelligent Engineering paradigm that will characterize future design systems, where rapid conceptualization and prototyping of engineering artifacts are the source of competitive edge.

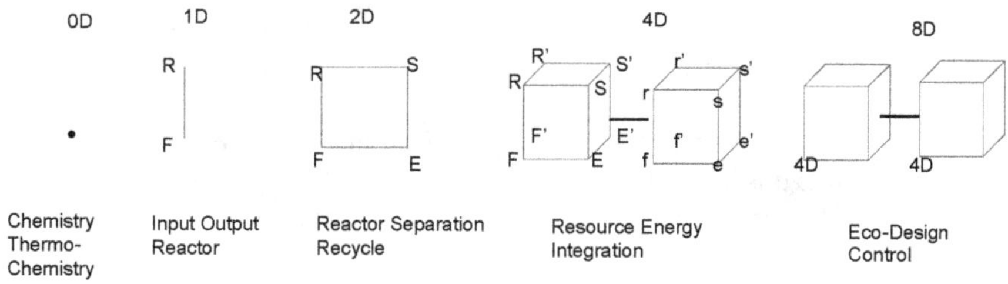

Fig. 6.11 Polytopic roadmap for conceptual design of chemical process

The Conceptual Design approach consists of a top-down analysis organized as a clearly defined sequence of tasks aggregated in levels, each one handling a fundamental conceptual problem. The considered Conceptual Design levels are:

(1) Chemistry and Thermo-chemistry
(2) Input/Output structure
(3) Reactor–Separation–Recycles
(4) Separation system
(5) Energy and resources integration
(6) Environment, safety and hazard problems
(7) Plant wide control

At any level, systematic methods may be applied for the analysis and synthesis of sublevels.

Figure 6.11 shows a Polytopic Roadmap for Conceptual Design of Chemical Process

The notations for Fig. 6.11 are:

- Feed-F
- Reactor-R
- Separation-S
- Exit-E

We use minuscule to denote control and apostrophes 'to denote dual or modified. The basic dual or modified modules are denoted as follows:

- Feed'-F'
- Reactor'-R'
- Separation'-S'
- Exit'-E'

The dimensions illustrated in Fig. 6.11 are:

- 0D-Chemistry Thermochemistry
- 1D-Input–Output. Reactor
- 2D-Reactor Separation Recycle
- 4D-Resource Energy Integration
- 8D-Eco-Design. Control

Observe that 0D includes level 1 of Conceptual Design. The 1D includes levels 2 and 3 of Conceptual Design. The 2D includes levels 2, 3 and 4. The 4D includes level 5.

The 8D includes levels 6 and 7 of Conceptual Design. Control refers to the second cube in 8D frame. It is a digital twin of the first cube in 8D.

With these allocations, the Douglas methodology can be considered as a precursor of the Intelligent Engineering paradigm.

Table 6.2 compares some Roadmaps for Chemical Engineering Paradigms

The stages or levels of the selected paradigms are reallocated according to the dimension of the associated reference architectures.

The presented roadmaps show partial similarity. It should be emphasized that the reference architecture dimension differs from scale but larger scales needs higher topological dimensions.

The key message is that we should think about the future paradigm frames as high dimensional.

When designing the paradigm, and the paths towards that paradigm, we are not looking towards a singular end-path. Rather, we are opening up space to fulfill many simultaneous and sometimes contradictory objectives. So, we need more than physical space, because future systems are not only reliant on 3D physical dimensions, but also include augmented and virtual dimensions of thinking. When we think of the future paradigms the mental map we need is not only 3D-cube but also includes several context conditions. The dimension we use does not describe spatial dimensions, but rather the topological dimension of the geometric objects we are describing. The assigned 4D or 8D reference architectures are still considered as included in the physical 3D space. This is valid for molecules, equipments, products and plants too.

The earlier paradigms as unit operations and transport phenomena remain significant in many practical engineering cases. We could admit that product engineering or sustainable chemical engineering, are the next paradigms, but more paradigms as intelligent chemical engineering should be anticipated along with industry further complexity growth.

The paradigms are important for engineering curriculum.

In terms of Dual Process Theory the paradigms are close to System 1, S1, while sciences are close to System 2, S2 thinking.

The scientific and engineering progress was studied in genetic epistemology (Piaget 1970; Piaget and Garcia 1989) and also in terms of paradigms and scientific revolutions (Kuhn 1996).

The complementarity of approaches of Kuhn (S1) and Piaget (S2) allow establishing the engineering curriculum (Tsou 2006). Sciences and paradigms should stand out as complementary rather than conflicting as inclusive rather then exclusive.

Table 6.3 shows a roadmap of chemical engineering paradigms and the associated sciences

Chemical engineering need an education system which delineates Kahn paradigmatic thinking, non-standard systems, S1, and activates Piaget's scientific, standard systems, S2, for specific curriculum on a case-by-case basis, depending on the education goals and context of the industry tasks.

Table 6.2 Roadmaps for chemical engineering paradigms

Dimension	0D	1D	2D	4D	8D
Conceptual design Douglas (1988)	Chemistry thermo-chemistry	Input–Output reactor	Reactor separator recycle	Resource energy integration	Eco-design control
Multiscale Villermaux (1996)	Chemistry	Operations	Product	Plant	Economy environment
Multiscale Charpentier (2010)	Nano molecule	Micro bubbles	Meso reactor separator	Macro plant	Mega environment
Polytopic roadmap	Industrial chemistry	Operationstransfer phenomena	Product	Sustainability self-evolvable industry 4.0	IntelligentSE of SE industry 8.0

Table 6.3 Chemical engineering paradigms and sciences

Dimension	0D	1D	2D	4D	8D
Paradigms S1	Industrial chemistry	Operations transfer phenomena	Product	Sustainability self-evolvableindustry 4.0	Intelligent SE of SE industry 8.0
Sciences S2	Chemistry	Chemical physics PDE	Product designDoE	Control, processsynthesis economy	Ecology artificial intelligence

SE-Self-Evolvable, PDE-Partial Differential Equations, DoE-Design of Experiments

Coupling new paradigms with up-to-date scientific and technical computations, chemical engineers can meet the challenges of high complexity projects.

Despite the fact that the presented roadmaps refer to chemical engineering the Polytopic Roadmaps methodologies based on reference architecture dimensionality aim to be general and equally applicable to other fields of engineering and science

References

Anderl, R., Picard, A., Wang, Y., Fleischer, J., Dosch, S., Klee, B., Bauer, J.: Guideline industry 4.0 - guiding principles for the implementation of industry 4.0 in small and medium sized businesses. In: VDMA Forum Industry 4.0, Frankfurt (2015)

Bird, R.B., Stewart, W.E., Lightfoot, E.N.: Transport Phenomena, 1st edn. J. Willey, New York (1960)

Charpentier, J.-C.: Perspective on multiscale methodology for product design and engineering. Comput. Chem. Eng.. Chem. Eng. **33**, 936–946 (2009)

Charpentier, J.-C.: Among the trends for a modern chemical engineering, the third paradigm: the time and length multiscale approach as an efficient tool for process intensification and product design and engineering. Chem. Eng. Res. Des. **88**(3), 248–254 (2010)

Clift, R.: Sustainable development and its implications for chemical engineering. Chem. Eng. Sci. **61**, 4179–4187 (2006)

Cussler, E.L., Moggridge, G.D.: Chemical Product Design, 1st edn. Cambridge University Press, New York (2001)

Douglas, J.M.: Conceptual Design of Chemical Processes. McGraw-Hill Inc., Singapore (1988)

Ehrenfeld, J.R.: Sustainability by Design: A Subversive Strategy for Transforming our Consumer Culture. Yale University Press, New Haven (2008)

Fischer, K.W.: A theory of cognitive development: the control and construction of hierarchies of skills. Psychol. Rev. **87**, 477–531 (1980)

Han, C., Stephanopoulos, G., Douglas, J.M.: Automation in design: the conceptual synthesis of chemical processing schemes. Adv. Chem. Eng. **21**, 93–146 (1995)

Heinzle, E., Biwer, A., Cooney, C.: Development of Sustainable Bioprocesses. J. Willey, West Sussex (2006)

Hill, M.: Chemical product engineering-the third paradigm. Comput. Chem. Eng.. Chem. Eng. **33**(5), 947–953 (2009)

Houdé, O.: 3-System Theory of the Cognitive Brain. A Post-Piagetian Approach to Cognitive Development. Routledge, London (2019)

Hougen, O., Watson, K.: Industrial Chemical Calculations, The Application of Physico-Chemical Principles and Data to Problems of Industry. John Wiley & Sons Inc., NY (1936)

Iordache, O.: Advanced Polytopic Projects. Springer, Cham (2019)

Kahneman, D.: Thinking, Fast and Slow. Farrar, Straus and Giroux, New York (2011)

Kuhn, T.S.: The Structure of Scientific Revolutions. University of Chicago Press, Chicago (1996)

Lerou, J.J., Ng, K.M.: Chemical reaction engineering: a multiscale approach to a multiobjective task. Chem. Eng. Sci. **51**, 1595–1614 (1996)

Li, J.H., Kwauk, M.S.: Particle-fluid Two-Phase flow: The Energy-Minimization Multi-scale Method. Metallurgical Industry Press, Beijing (1994)

Li, J.H., Kwauk, M.S.: Exploring complex systems in chemical engineering—the multiscale methodology, Chem. Eng. Sci. **58**(3–6), 521–535 (2003)

Lupasco, S.: Le principe d'antagonisme et la logique de l'énergie. Prolégomènes à une science de la contradiction. Hermann & Co., Paris (1951)

Narodoslawsky, M.: Chemical engineering in a sustainable economy. Chem. Eng. Res. Des. **91**(10), 2021–2028 (2013)

Piaget, J.: Genetic Epistemology. Columbia University Press, New York (1970)
Piaget, J., Garcia, R.: Psychogenesis and the History of Science. Columbia University Press, New York (1989)
Stephanopoulos, G.: Artificial intelligence in process engineering: a research program. Chem. Eng. Educ. 182–185 (1986)
Stock, T., Seliger, G.: Opportunities of sustainable manufacturing in industry 4.0. In: Procedia CIRP, vol. 40, pp. 536–541 (2016)
Thiel, T.: The design of the connection machine. Des. Issues **10**, 1, 5–18 (1994)
Tsou, J.Y.: Genetic epistemology and Piaget's philosophy of science. Piaget vs. Kuhn on scientific progress. Theory & Psychol. **16**, 203–224 (2006)
Uslar, M., Engel, D.: Towards generic domain reference designation: how to learn from smart grid interoperability. In: Poster Proceedings of DACH Energy Informatics 2015, pp. 1–12 (2015)
Uslar, M., Gottschalk M.: Extending the SGAM for electric vehicles. In: International ETG Congress 2015; Die Energiewende-Blueprints for the New Energy Age; Proceedings of 2015 Nov 17, pp. 1–8 (2015)
Uslar, M., Trefke, J.: Applying the smart grid architecture model SGAM to the EV domain. In: EnviroInfo 2014, pp. 821–826 (2014)
Villermaux, J.: New horizons in chemical engineering. In: Summary proceedings 5th World Congress of Chemical Engineering, 14–18 July 1966, pp. 16–23. A.I.Ch.E., San Diego, California, USA (1966)
Walker, H., Lewis, W.K., Mc Adams, W.H.: Principles of Chemical Engineering. Mc Graw-Hill, New York (1923)

Chapter 7
Systems

Abstract The relation between high complexity and systems theory is delineated. An architecture framework for systems of systems, SoSAF, is presented. Applicability to manufacturing production systems is evaluated. The process of making good assessments reduces to decision making models. For command and control applications during an actual operation, the opportunity time is limited and uncertainty is high. The 8D development of the OODA (observe-orient-decide-act) model is introduced. Operations process and cyber-physical social system are analyzed in terms of the general 8D framework.

7.1 Production Systems

The dynamics of modern markets require the development of new product and production processes to be carried out side by side. However, in the production systems the interactions and their relationships among many variables are not known due to increasing complexity. Thanks to adopting just-in-time and concurrent strategies, even small perturbations can cause chain effects exacerbated by the enhanced dynamicity and connectivity involved along the processes. A more complex organization requires more formal procedures like architecture to comprehend complexities and to have an effective control over production flows and capacities. Without an appropriate architecture, the systems are just piled together so they may conflict with each other.

Architecture formation is an important part of SoS design process. The design of the SoS requires that analysis methods are appropriate to the type of entities that constitute the system of systems (DeLaurentis and Crossley 2005). Production SoS design process comprises different phases starting from idea generation, strategy formulation, selection and architecture modeling. In order to map such architecture, the designing and execution of a Production System architecture based on System of Systems Approach—named SoSAF (System of Systems Architecture Frameswork)—has been introduced (Mahmood and Montagna 2012).

O. Iordache, *General Reference Architecture Frameworks*, Studies in Systems, Decision and Control 73, https://doi.org/10.1007/978-3-031-70718-6_7

Having such architecture is a precondition for well-informed decisions because it provides a comprehensive view of the processes and acts like a regulator. It should avoid, for example, a maintenance system to be developed in isolation from manufacturing, quality and scheduling; rather it would facilitate to evolve in an interactive and collaborative environment with other systems. In SoSAF, the net centric environment would establish communications across collaborating systems and operate with the network instead of point to point communications. SoSAF is a general architectural framework yet to be constructed that can have inspirations from previously developed governmental and non-governmental architectures.

The production systems are heterogeneous systems because they perform different functional activities in the production SoS, employ different resources of hardware and software, interact differently abroad the organization (suppliers and customers). The communication, cooperation and collaboration among these heterogeneous systems are the essential purposes of production architecture.

To achieve this, multiple systems are required to be simultaneously connected and there is a need to delineate the over-all behavior of the SoS as a whole.

For architecture designing a reference framework and a modeling language are needed (DoDAF 2009).

For production SoS, there are various architectures available that could potentially act as reference, for example, some military architecture are MODAF (UK), NAF (NATO), DNDAF (Canada), MDAF (Italy), AGATE (France) and ADOAF (Australia).

DoDAF (defense) can be adapted to civil production architectures because it can distinctively be used to design network of systems where communication is the basic element for the successful completion of mission—the strategic goal to face a scenario in the production case- and because various multiple perspectives can be extracted as views of the whole production architecture (DoDAF 2009).

Different problems in architectural frameworks may require a single or multi-viewpoints with different levels of details. Each representation caters specific objectives and engenders different effects on the overall system operation.

The production SoSAF can be arranged into 4 viewpoints with 4 perspectives for each viewpoint.

Manufacturing Viewpoint (M) is the viewpoint around which other viewpoints are grounded. It adds the value by converting raw materials and components into end products. Quality Viewpoint (Q) is the description of procedures, and a total quality culture. It ensures the quality program as an integral part of the strategic objectives of the company, extending their values to the other viewpoints.

Technical Maintenance Viewpoint (T) describes the functions and readiness of the shop floor. It keeps the maintenance data and past solutions in historical data servers to act as input data for future applications. It is simpler for an operator to decode the problem when corrective maintenance action is connected with errors. The data flows at different terminals and stores in a common database in the form of history of equipment condition, cause and frequency of disruptions.

Production Planning and Scheduling Viewpoint (P) is correlated and two sequential integrated processes, and therefore, should be dealt likewise. To ensure the optimum utilization of resources and shortest production lead time, short, medium and alternative plans are required. It issues operational orders to the machines, supervise and coordinate the execution process, direct routing and equipment selection. It is the complex and dynamic activity due to numerous variables involved and unanticipated events.

There have to be clear relationships and communications among these viewpoints.

These viewpoints provide insight of objectives, capabilities, constituent description and enable to slice the problem in a specialized fashion.

Each viewpoint has 4 perspectives that are:

- Systems and Services Perspective (S) supports operational activities by providing system functions and service resources. It provides services in the form of a series of processes to keep production process an ongoing phenomenon
- Operational Perspective (O) describes the activities, operational nodes and the information exchanges to complete SoS missions
- Technical Standard Perspective (T) provides the set of rules, criteria, technical standards for operational activities
- Application Perspective (A) as a fourth perspective is introduced

The set of perspectives is denoted SOTA.

Every viewpoint is described by these four perspectives, for example, M viewpoint is described: by SOTA related to the machineries, equipment in practice (technical standard perspective) by managerial practices (operational perspective), and: by flow charts of machines and material, automation and other means to secure smooth operation (systems and services perspective).

Figure 7.1 shows the Polytopic Roadmap for SoS.

In the 8D frame each step of 4D frame is replaced by 4D component sub-steps.

The notations are:

- Planning-P
- Manufacturing-M
- Quality-Q
- Technical Maintenance-T

The dimensions illustrated in Fig. 7.1 are:

- 0D-Planning

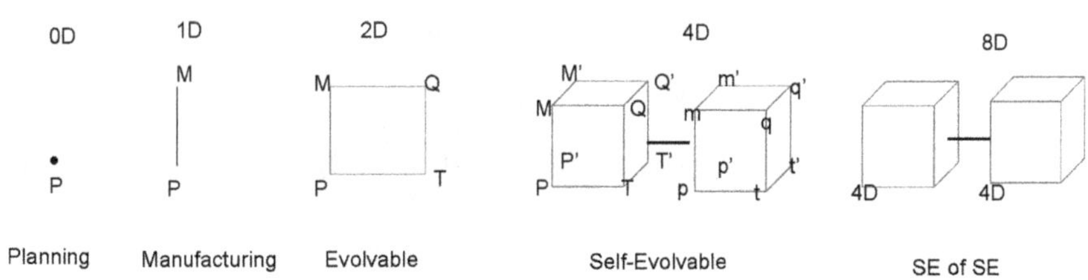

Fig. 7.1 Polytopic roadmap for SoS

- 1D-Manufacturing
- 2D-Evolvable
- 4D-Self-Evolvable
- 8D-SE of SE

We use the minuscule p, m, r, t to denote the control of Direct sequence P, M, Q, T and the minuscule p', m', r', t' to denote the control of Reverse sequence P', M', Q', T'.

The 4D polytope shown in Fig. 7.1 illustrates the architecture of potentially self-evolvable systems. It is a representation consisting from a pair of 3D-cubes. The 4D-cube is obtained by joining all corners of the first 3D-cube with the corresponding corners of the second 3D-cube. The second cube is the digital twin of the first cube.

The 8 lines connecting the two 3D-cubes are represented in Fig. 7.1 by a thick hyper-line. This representation is useful for 4D and 8D polytope too. For 8D a thick hyper-line connects the two 7D-cubes generating the 8D polytope (Thiel 1994).

Figure 7.2 shows more details and another presentation for the polytope 4D for SoS presented in the Fig. 7.1.stage 4D (Iordache 2013, 2017).

The notations associated to Self or control are:

- p-Planning control
- m-Manufacturing control
- q-Quality control
- t-Technical Maintenance control

The polytope 4D outlines that in interaction with the converging, constructing or Processing way P → M → Q → T we need to look at the diverging, deconstructing or Reprocessing way T′ → Q′ → M′ → P′.

The swinging from the Direct way of Processing to the dual Reverse way of Reprocessing is critical for coexisting tendencies of unification and diversification.

Fig. 7.2 Polytope 4D for SoS

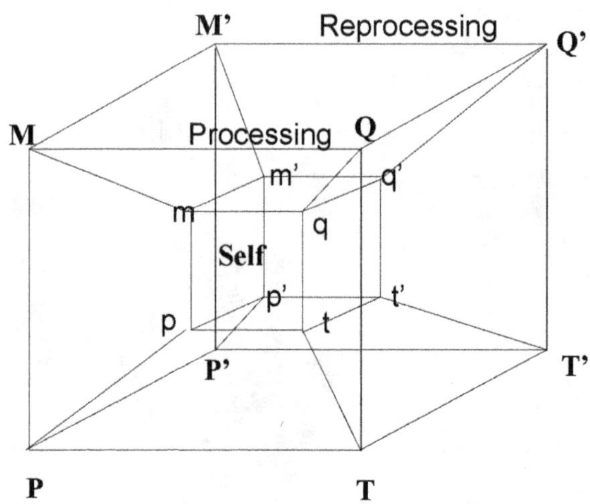

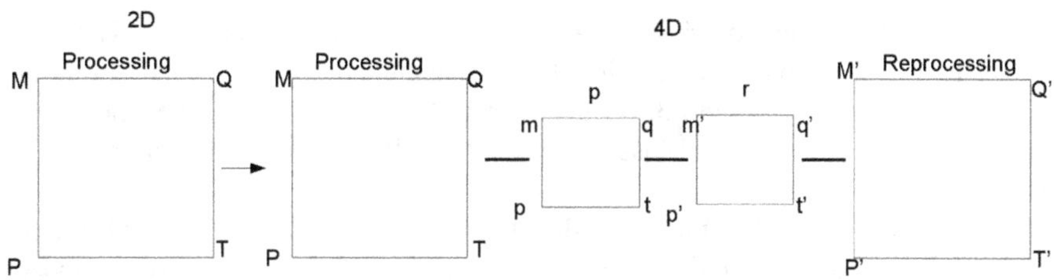

Fig. 7.3 Developed roadmap for polytope 4D for SoS

Since dual complementary descriptions are intrinsically irreducible and in some cases contradictory, the concept integration or coordination of complementary descriptions requires a higher level description, a new dimension that emerges as a new hierarchical level of complexity. It is the 4D and beyond reference architecture.

The inner cube, marked as the Self allows dialogue and interaction of the two ways in duality relation. These ways correspond to construction or Direct way of Processing, from P to T, or to deconstruction or Reverse way of Reprocessing from T′ to P′.

The Self needs to endorse the dialogue, to ensure coordination and coherence of evolution on different levels. It results from Fig. 7.2 that the structure of the inner cube is isomorphic to that of the outer cube. Observe that the two cubes of 4D-cube are reunited in a compact way. The 4D-cube may be considered as a two scale frame. The inner cube is a digital twin of the outer cube.

After any step in the Direct sequence of Processing, the systems make contact with the Self represented by the inner cube. Signals as p, m, q and t reinforce the evolution according the Direct Processing sequence. If the Self have to change the direction shifting to t', q'.m', p' the next step will be along the Reverse sequence of Reprocessing.

The set (p', m', q', t') corresponds to the control of Reverse sequence.

Figure 7.3 shows a developed roadmap for polytope 4D for SoS.

The 2D and 4D frames are presented in a developed form here.

The 2D frames are based on Processing. Processing implies Reprocessing.

In this step a new frame called the Self or Included Middle is considered.

The Self is shown with Processing p, part and Reprocessing r, part.

Since Processing and Reprocessing frames are 2D the p and r frames are 2D too.

Relating p and r allows obtaining the inner cube. Finally the 4D frame results, by inclusion of the inner cubic frame in the outer cubic frame. It results a 4D-cube representing the self-evolvable, SE systems. The inner cube is a digital twin of the outer cube.

The developed roadmap may be interpreted in terms of dual process theory (Kahneman 2011). The Processing P → M → Q → T pertains to System 1 while the Reprocessing T′ → Q′ → M′ → P′ pertains to System 2.

Lupasco (1951) and Houdé (2019) go beyond the approaches based on System 1, and System 2, to develop a theory in which the included third T, or the inhibitory-control system that is System 3, S3 takes precedence. T or S3 denotes the bridge between S1 and S2. Here S3 is associated to the Self and links the p and r parts of IM.

The grafting strategy that maintains the 4D architecture but replace any 0D vertices by developed and structured 4D vertices is considered. It gives a hypercube of hypercubes.

The new dimension is 8D = 4D + 4D. The first 4D come from the initial 4D frame as shown in Fig. 7.2 and the next 4D from the component 4D frames replacing the vertices of the initial hypercube. Hypercubes as P refers to stages while hypercubes as p refers to control. Figure 7.4 shows the polytope 8D for SoS.

This is the GRAF developed in the 8D Program.

The presented Roadmaps outlined two sequences for evolution the direct sequence: P → M → Q → T that corresponds to the Processing stages and the reverse sequence T′ → Q′ → M′ → P′ that corresponds to the Reprocessing stages.

It results a sustainable type of development. The directly acquired development is questioned and modified by the reverse sequence. The bottom-up construction from direct sequence complements the top-down deconstruction of the reverse sequence.

At any stage there are 4 sub-stages called perspectives and denoted also by:

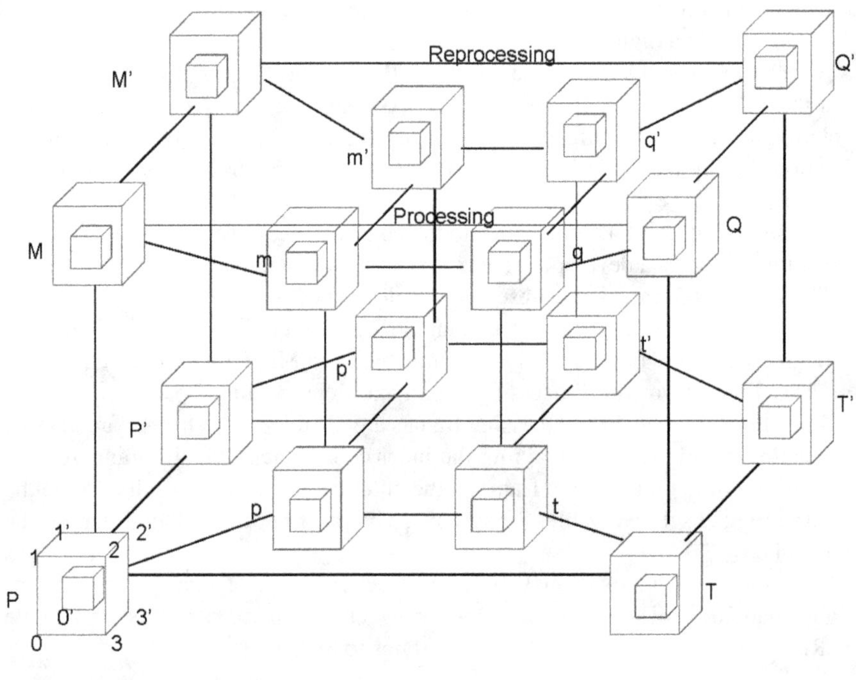

Fig. 7.4 Polytope 8D for SoS

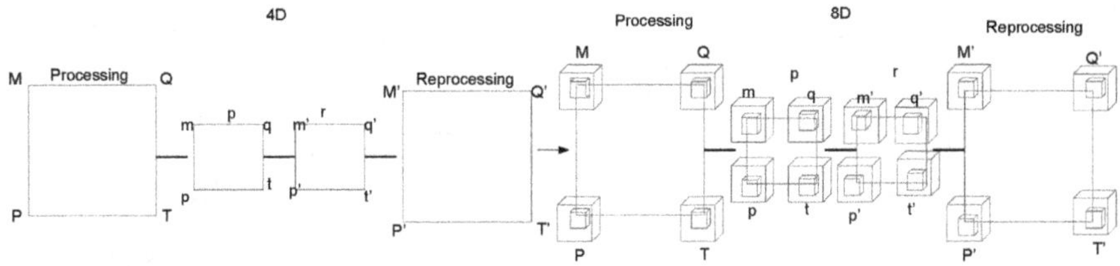

Fig. 7.5 Developed roadmap for polytope 8D for SoS

- 0-Systems-S
- 1-Operatorial-O
- 2-Technology-T
- 3-Application-A

For dual sequence the notations are:

- 0'-Systems-S'
- 1'-Operatorial-O'
- 2'-Technology-T'
- 3'-Application-A'

Apostrophes 'mark the dual or modified states.

Figure 7.5 shows a developed roadmap for polytope 8D for SoS.

The 4D and 8D frames are presented here.

The initial 4D architecture shows the developed frame of an external and an internal cube. These cubes result by relating the corresponding 2D frames.

The transition from 4 to 8D is achieved by replacing any 0D vertex from the initial 4D architecture by component 4D frames. It is a strategy that keeps the initial 4D architecture but replace any 0D vertices by component 4D vertices.

It results a hypercube of hypercubes. The new dimension is 8D = 4D + 4D. The first 4D come from the initial 4D frame and the next 4D from the component 4D frames replacing the 0D dots of the initial 4D hypercube

In the next step an inclusion of a new frame called Self is considered. The Self is shown with Processing p, part and Reprocessing r, part.

Relating p and r allows obtaining the inner 7D-cube. Finally the 8D frame results, by inclusion of the inner 7D-cube in the external 7D-ube.

The inner 7D-cube is a digital twin of the outer 7D-cube.

7.2 Decision Making

Decision making is at the core of human cognitive activity. The process of making good decisions is a long studied skill that has drawn from previous human experience and been condensed into a number of decision making models. Many decision making models share common aspects and attributes but differ in the order, area of

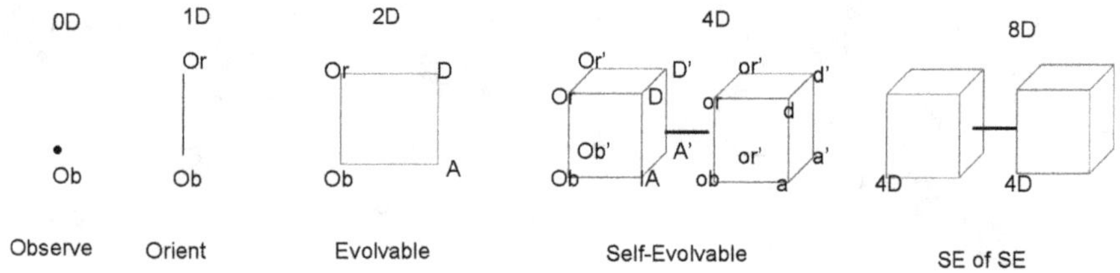

Fig. 7.6 Polytopic roadmap for OODA

emphasis, or underlying assumptions. For situations where there is ample time to consider and analyze the situation before making a decision, analytical models can be considered appropriate. However, for command and control applications during an actual operation, time is limited and uncertainty is high.

The OODA loop is the cycle observe–orient–decide–act, developed by strategists (Boyd 2007). They applied the concept to the operations process, often at the operational level (Richards 2004; Dreier 2012).

It is now often applied to understand commercial operations and learning processes.

The approach explains how agility can overcome raw power in dealing with human opponents. It is especially applicable to cyber security and cyber warfare.

The OODA loop has become an important concept in litigation, business, law enforcement, management education, and strategy. Decision occurs in a recurring cycle of observe–orient–decide–act. An entity, whether an individual or an organization, that can process this cycle quickly, observing and reacting to unfolding events more rapidly than an opponent, can thereby get inside the opponent's decision cycle and gain the advantage.

Figure 7.6 shows the Polytopic Roadmap for OODA.

The notations are:

- Observe-Ob
- Orient-Or
- Decide-D
- Action-A

The dimensions illustrated in Fig. 7.6 are:

- 0D-Observe
- 1D-Orient
- 2D-Evolvable
- 4D-Self-Evolvable
- 8D-SE of SE

We use the minuscule ob, or, d, a to denote the control of Direct sequence and the minuscule ob', or', d', t' to denote the control of Reverse sequence.

The 4D polytope shown in Fig. 7.6 illustrates the architecture of potentially self-evolvable systems. It is a representation consisting from a pair of 3D-cubes.

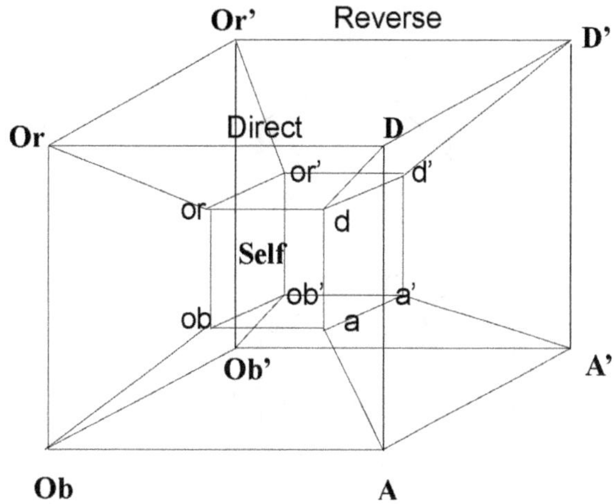

Fig. 7.7 Polytope 4D for OODA

The second cube is the digital twin of the first cube in the 4D frame.

The 4D-cube is obtained by joining all corners of the first 3D-cube with the corresponding corners of the second 3D-cube. The 8 lines connecting the two 3D-cubes are represented in Fig. 7.6 by a thick hyper-line. This representation is useful for 8D polytope too.

Figure 7.7 shows a different presentation of the Polytope 4D for OODA.

Figure 7.8 shows a developed roadmap for polytope 4D for OODA.

The 2D and 4D frames are presented in a developed form here.

The 2D frames are based on Direct sequence. This implies Reverse sequence.

Then a new frame the Self is considered.

The Self is shown with Direct d, part and Reverse r, part.

Since Direct and Reverse frames are 2D the d and r frames are 2D too. They illustrate the conditions. Relating d and r allows obtaining the inner cube. Finally the 4D frame results, by inclusion of the inner cube in the outer cube. It will be a 4D-cube representing the self-evolvable systems. The inner cube is a digital twin of the outer cube.

The developed roadmap may be interpreted in terms of dual process theory (Kahneman 2011). The Direct sequence Ob → Or → D → A pertains to System 1

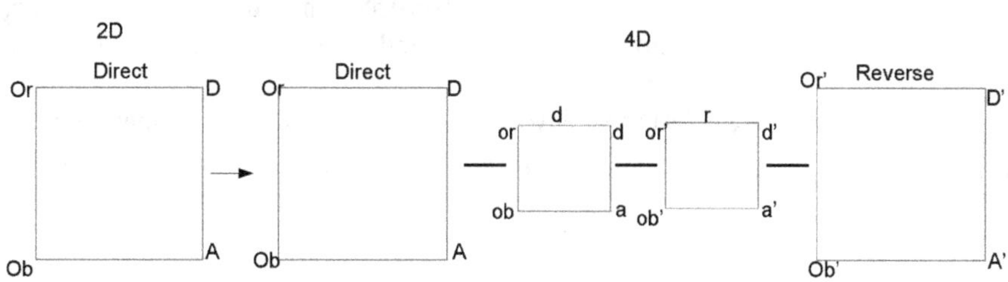

Fig. 7.8 Developed roadmap for polytope 4D for OODA

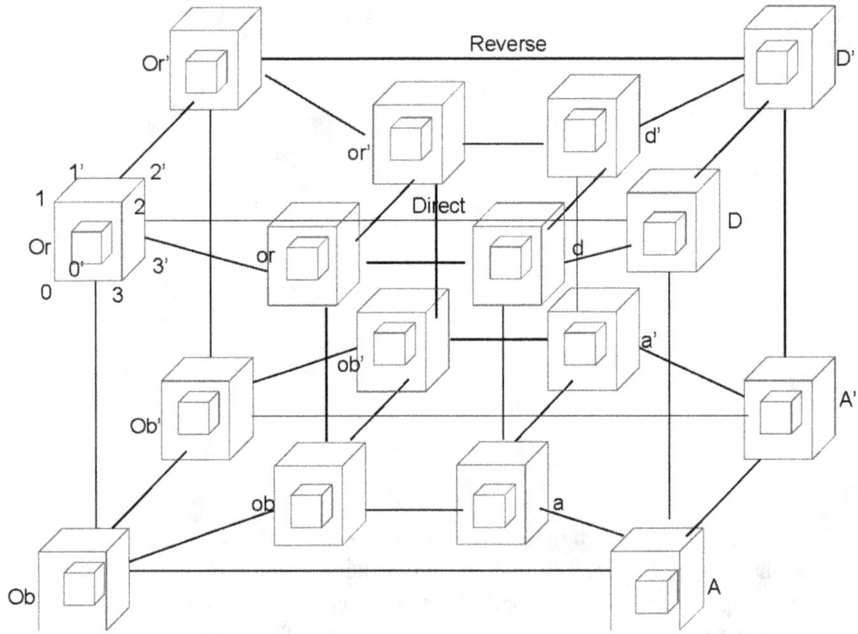

Fig. 7.9 Polytope 8D for OODA

while the Reverse sequence A′ → D′ → Or′ → Ob′ pertains to System 2. Apostrophes 'mark the dual states.

A theory in which the included third T, that is the inhibitory-control system System 3, S3 take precedence was developed. T or S3 denote the bridge between S1 and S2 (Lupasco 1951; Houdé 2019).

Here S3 is associated to the Self putting together the d and r parts. S3 is a 3D-Cube.

Figure 7.9 shows a different presentation of the Polytope 8D for OODA.

The presented Roadmaps outlined two sequences for evolution the direct sequence: Ob → Or → D → A that corresponds to the Direct stages and the reverse sequence A′ → D′ → Or′ → Ob′ that corresponds to the Reverse stages.

This is a sustainable type of development. The directly acquired development is questioned and modified by the reverse sequence. The bottom-up construction from direct sequence complements the top-down deconstruction of the reverse sequence.

For any levels we may continue experiments and their analysis and synthesis in sublevels.

Following Richards (2004) as sublevels of the level or modules we may refer to:

- 0-previous expertise
- 1-new info
- 2-analysis
- 3-synthesis

Apostrophes 'mark the dual or modified states.

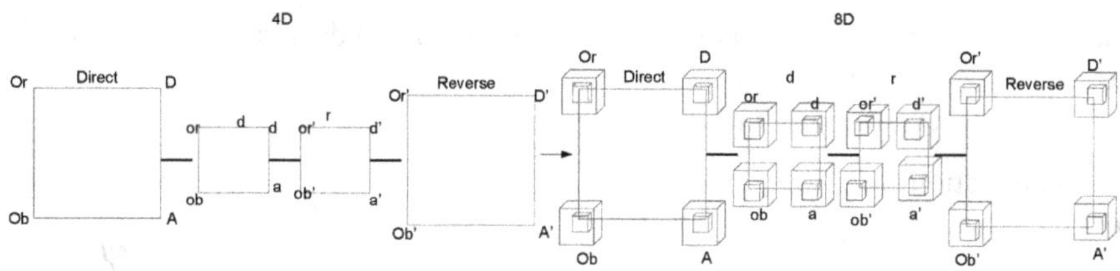

Fig. 7.10 Developed roadmap for polytope 8D for OODA

For dual sequence the notations are:

- 0'-previous expertise'
- 1'-new info'
- 2'-analysis'
- 3'-synthesis'

Notice that Richards (2004) refers to Orient module only.

Figure 7.10 shows a developed roadmap for polytope 8D for OODA.

The 4D and 8D frames are presented here.

The initial 4D architecture shows the developed frame that includes an external and an internal cube. These cubes result by relating the corresponding 2D frames.

The transition from 4 to 8D is achieved by replacing any 0D vertex from the initial 4D architecture by component 4D frames. It is a grafting strategy that keeps the initial 4D architecture but replace any 0D vertices by component 4D vertices.

It results a hypercube of hypercubes. The new dimension is 8D = 4D + 4D. The first 4D come from the initial 4D frame and the next 4D from the component 4D frames replacing the 0D dots of the initial 4D hypercube

In the next step an inclusion of a new frame called Self is considered. The Self is shown with Direct d, part and Reverse r, part.

Relating d and r allows obtaining the inner 7D-Cube. Finally the 8D frame results, by inclusion of the inner 7D-cube in the outer 7D-Cube. The inner 7D-cube is the digital twin of the outer 7D-cube.

Different studies of higher dimensionality for OODA loops are due to Gustavsson and Planstedt (2005).

7.3 Operations Process

An operations process is defined as the organizational method for getting a task accomplished.

There exist different approaches for operations processes.

Marshall (1995) and Paris et al (1998) have been interested in exploring the way individuals and organizations made decisions in the field. They abandoned the traditional way of defining knowledge as procedural and declarative and defined a new set of possible operations.

The considered actions have been:

- Identification
- Planning
- Elaboration
- Execution

They correspond to the knowledge types and use.

Identification is used in the creation of pattern recognition.

Planning is used to connect a goal state to a set of possible actions to realize that state.

Elaboration is used in the creation of a mental model about the current problem situation.

Execution is used to guide implementation and determine who should perform required actions.

In what follows it is considered that operation process consists of four distinct primary activities:

- Planning
- Preparing
- Executing
- Assessing

This definition is valid for any type of operations, civil or military (Papanastasiou 2009).

The framework for exercising command is the so-called operations process. Leaders, use the operations process to drive the conceptual and detailed planning necessary to understand, visualize, and describe their operational environment, make and articulate decisions, and direct, lead, and assess operations (Smith 2005; Papanastasiou 2009).

The operations process constitutes the leaders view on planning, preparing, executing and assessing operations.

It accounts for the complex, ever-changing, and uncertain nature of operations and recognizes that an operation is foremost a human undertaking. It is a need to emphasize the values of mission command to include the central role of leaders, supported by the team, in driving the operations process.

Planning starts an iteration of the operations process. Upon completion of the initial order, planning continues as leaders revise the plan based on changing circumstances. Preparing begins during planning and continues through execution. Execution puts a plan into action by applying enterprise power to seize, retain, and exploit the initiative to gain a position of relative advantage.

Planning is the art and science of understanding a situation, envisioning a desired future, and laying out effective ways of bringing that future about. Leaders plan to create a common vision among teams, staffs, and unified action partners for the successful execution of operations. Planning results in a plan or order that communicates this vision and directs actions to synchronize forces in time, space, and purpose for achieving objectives and accomplishing missions.

Time is a critical variable in operations. Therefore, time management is important in planning. Whether done deliberately or rapidly, all planning requires the skilled use of available time to optimize planning and preparation throughout the operating unit.

Taking more time to plan often results in greater synchronization; however, any delay in execution risks yielding the initiative-with more time to prepare and act-to the competitors. When allocating planning time to the staff, leaders must ensure assistants have enough time to plan and prepare their own actions prior to execution.

Preparation consists of those activities performed by units and operators to improve their ability to execute an operation. Preparation creates conditions that improve working forces' opportunities for success. Mission success depends as much on preparation as on planning. Higher command centers may develop the best of plans; however, plans serve little purpose if subordinates do not receive them in time. Subordinates need enough time to fully comprehend the plan, and rehearse key portions of the plan.

Operators and equipment should be positioned and ready to execute the operation.

Execution is putting a plan into action by applying working power to accomplish the mission. During execution, leaders, staffs, and teams commanders focus their efforts on translating decisions into actions. They apply working power to seize, retain, and exploit the initiative to gain and maintain a position of relative advantage. This is the essence of unified operations.

Assessment is the determination of the progress toward accomplishing a task, creating an effect, or achieving an objective. Assessment is a continuous activity of the operations process. The focus of assessment, however, changes for each operations process activity. During planning, assessment focuses on understanding current conditions of an operational environment and developing an assessment plan, including what and how to assess progress. During preparation, assessment focuses on determining the force's readiness to execute the operation and on verifying the assumptions on which the plan is based. During execution, assessment focuses on evaluating progress of the operation. Based on their assessment, leaders direct adjustments, ensuring the operation stays focused on accomplishing the mission. The activities of the operations process are not discrete; they overlap and recur as circumstances demand.

Figure 7.11 shows the Polytopic Roadmap for operations process

The notations are:

- Plan-Pl
- Prepare-Pr
- Execute-Ex
- Assesses-As

The dimensions illustrated in Fig. 7.11 are:

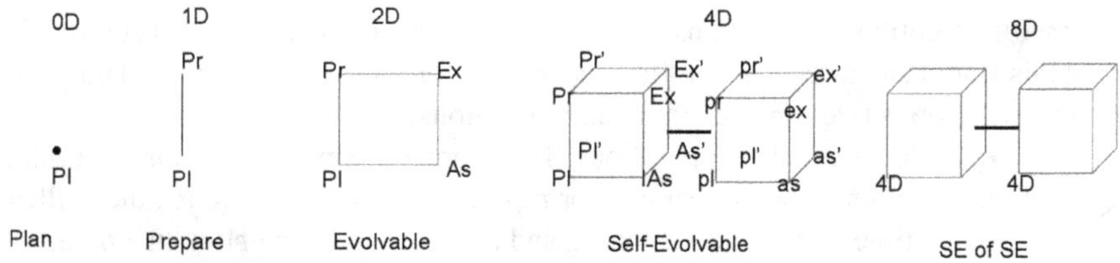

Fig. 7.11 Polytopic roadmap for operations process

- 0D-Plan
- 1D-Prepare
- 2D-Evolvable
- 4D-Self-Evolvable
- 8D-SE of SE

We use the minuscule pl, pr, ex, as to denote the control of Direct sequence and the minuscule pl', pr', ex', as' to denote the control of Reverse sequence.

The 4D polytope shown in Fig. 7.11 illustrates the architecture of potentially self-evolvable systems. It is a representation consisting from a pair of 3D-cubes.

The second cube is a digital twin of the first cube. The 4D-cube is obtained by joining all corners of the first 3D-cube with the corresponding corners of the second 3D-cube. The 8 lines connecting the two 3D-cubes are represented in Fig. 7.11 by a thick hyper-line. In this way the 4D-cube has a simplified representation. This representation is useful for 8D polytope too.

Figure 7.12 shows a different presentation for the Polytope 4D for operations process.

The inner cube is a digital twin of the outer cube.

Figure 7.13 shows a developed roadmap for polytope 4D for operations process.

The 2D and 4D frames are presented in a developed form here.

Fig. 7.12 Polytope 4D for operations process

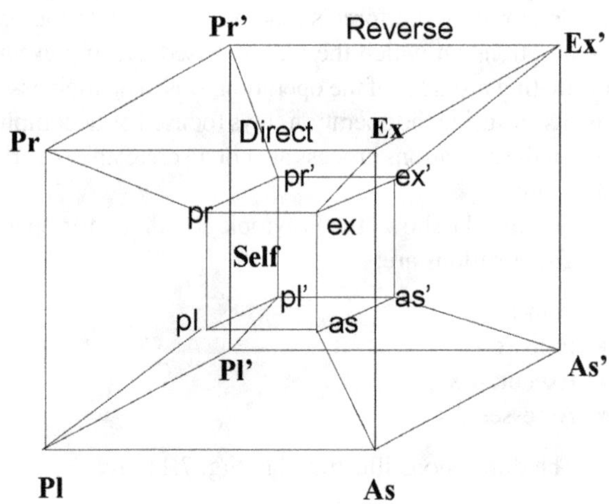

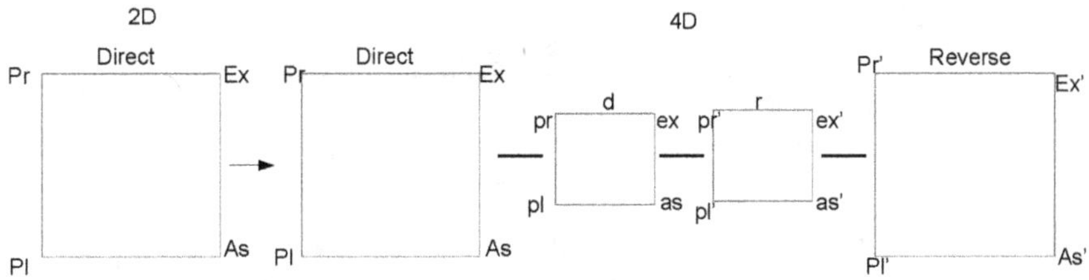

Fig. 7.13 Developed roadmap for polytope 4D for operations process

The 2D frames are based on operation sequences are considered as 2D frame. Direct sequence imposes Reverse sequence.

In the next step the new frame called the Self should be considered.

The Self is shown with Direct d, part and Reverse r, part. Since Direct and Reverse frames are 2D the d and r frames should be 2D too. They may illustrate the conditions.

Relating e and r allows obtaining the inner cube.

The 4D frame results by inclusion of the inner cube frame in the outer cube frame. It will be a 4D-cube representing the Self-Evolvable (SE) systems. The inner cube is a digital twin of the outer cube.

The new presentation outlines the isomorphism between the operations process and their controls.

Figure 7.14 shows a different presentation of the Polytope 8D for operations process.

As sublevels of the main levels we may refer to:

- 0-understanding
- 1-decision
- 2-controling
- 3-assesing

Apostrophes 'mark the dual states.

The dual sublevels are:

- 0'-understanding'
- 1'-decision'
- 2'-controling'
- 3'-assesing'

Figure 7.15 shows the roads to higher dimensional operations process.

S1 and S2 roads have been outlined.

Consider that the S2 road starts from Plan, 0D, represented here as a circle. Included in this an S1 frame appears. It is represented as an included dot. The dot and the circle will be related in a segment or line, 1D, as shown on the S2 road. The segment corresponds to Prepare.

Then on the S1 road an included middle represented as a segment appears. Transition to S2 road imposes the development of the segments. By relating the two segments the 2D square results. The square is associated to Evolvable stage.

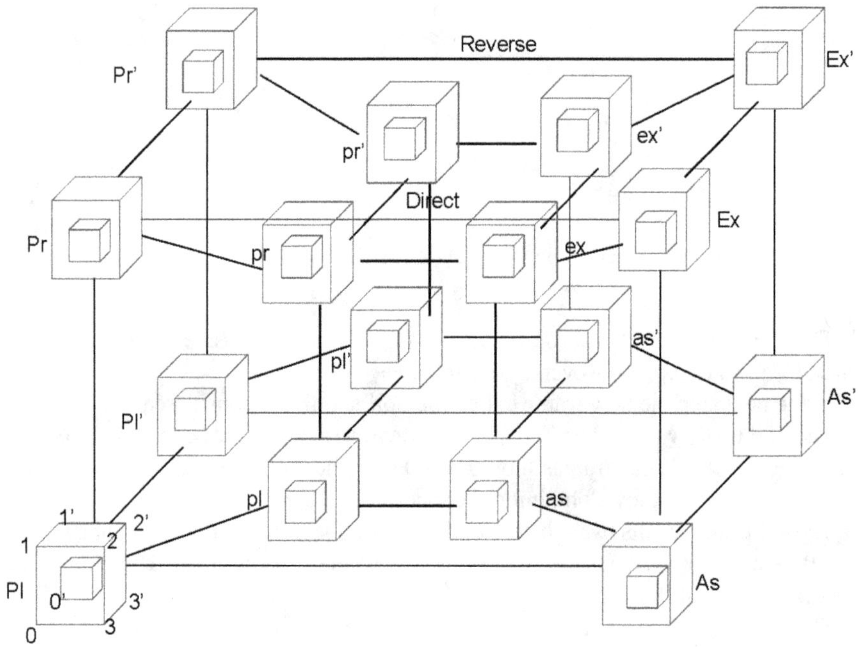

Fig. 7.14 Polytope 8D for operations process

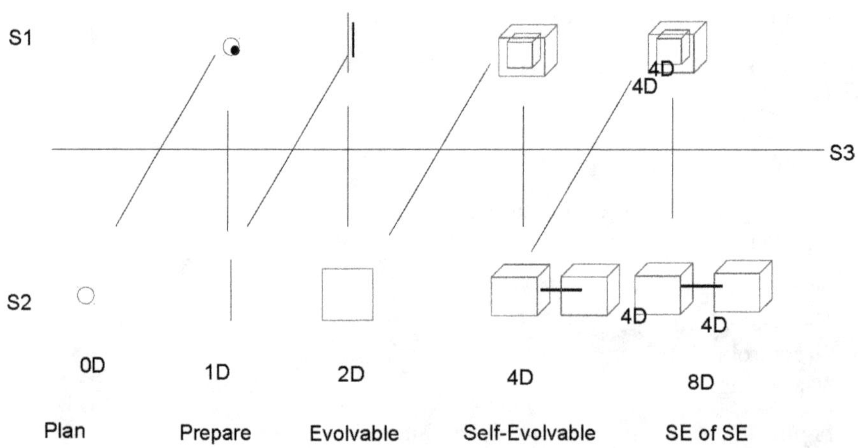

Fig. 7.15 Roads to higher dimensional operations process

Then on the S1 road a cube and an included cube appear and develop as a 4D hypercube. Transition to S2 road imposes the development and the correlation of the two cubes. The 4D–cube is resulting. It is associated to Self-Evolvable stage.

The 4D-cube on the S2 road is obtained by joining all corners of the first 3D-cube with the corresponding corners of the second 3D-cube. The 8 lines connecting the two 3D-cubes are represented in Fig. 7.15 by a thick hyper-line. In this way the 4D-cube has a simplified representation. This representation is useful for 8D polytope too.

The framework 8D is resulting as 4D of 4D architecture.

The process continues indefinitely in the same way as a zigzag between S1 and S2 roads.

7.4 Cyber Physical Social Systems

Cyber-physical systems (CPS) are smart systems that have cyber technologies, both hardware and software, deeply embedded in and interacting with physical components, sensing and changing the state of the real world. They represent a core opportunity area and source of competitive advantage for the innovation economy.

Examples of CPS-based initiatives include, smart factory, smart transportation systems, smart medical devices and technologies, smart buildings, next-generation air transportation systems, Smart Grid (Baheti and Gill 2000, Fang et al. 2012) and so on.

For CPS the real (physical) objects and processes associated with information processing (virtual) objects and processes are anytime interconnected information networks.

A cyber-physical-social system, CPSS, considers human factors as a part of a system instead of placing them outside the system boundary (Wang 2010).

Such systems consist of not only cyberspace and physical space but also human knowledge, mental capabilities and socio cultural elements.

Information from cyberspace interacts with physical and mental spaces in the real world as well as the artificial space mapping different facets of the real world.

The CPSS carries out parallel execution, self-synchronization, and influences in the physical, information cognitive and social domains.

These four domains are also the key elements in the command and control systems.

The efficiency of a command and control organization stems from the fusion of the human factors.

The CPSS provide an ideal paradigm for the design and construction of command and control organizations (Liu et al. 2011).

A Polytopic Roadmap for command and control is shown if Fig. 7.16.

The dimensions illustrated in Fig. 7.16 are:

- 0D-Physical
- 1D-Cognitive
- 2D-Evolvable
- 4D-Self-Evolvable
- 8D-SE of SE

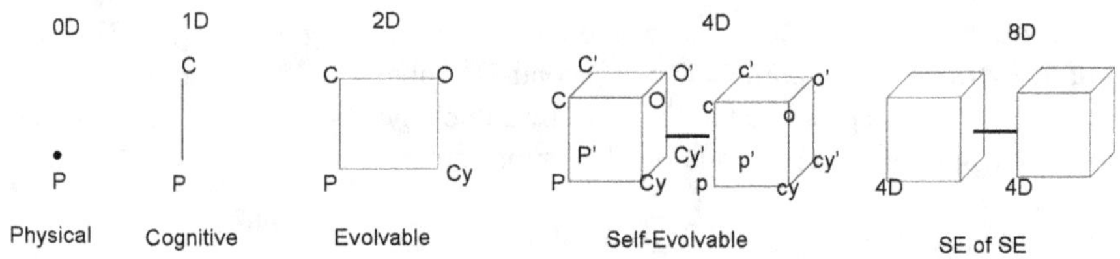

Fig. 7.16 Polytopic roadmap for command and control

The main stages of organization and their abbreviations are:

- K0-Physical-P
- K1-Cognitive-C
- K2-Organizational-O
- K3-Cyber-Cy

K0 corresponds to artifacts networks, K1 to a cognitive network, K2 to an organizational coordination network and K3 to the cyber network.

Sub-stages may be considered for any of these four main stages.

For instance for the stage K0 we may observe the sub-stages (Liu et al 2011):

- 0-devices
- 1-sensors
- 2-enablers
- 3-communications

These parallel the sequence 0-singles, 1-mappings, 2-systems, 3-SoS (Fischer 2008).

The sub-stages 0', 1', 2', 3'denotes modifications or improvements.

- 0'-devices'
- 1'-sensors'
- 2'-enablers'
- 3'-communications'

We use the minuscule p, c, o, cy to denote the control of Direct sequence and the minuscule p', c', o', cy' to denote the control of Reverse sequence.

The 4D polytope shown in Fig. 7.16 illustrates the architecture of potentially self-evolvable systems. It is a representation consisting from a pair of 3D-cubes.

The second cube is a digital twin of the first cube.

The 4D-cube is obtained by joining all corners of the first 3D-cube with the corresponding corners of the second 3D-cube. The 8 lines connecting the two 3D-cubes are represented in Fig. 7.16 by a thick hyper-line. This representation is useful for 8D polytope too.

Figure 7.17 shows the polytope 8D for developed operational mechanism including modifications and improvements.

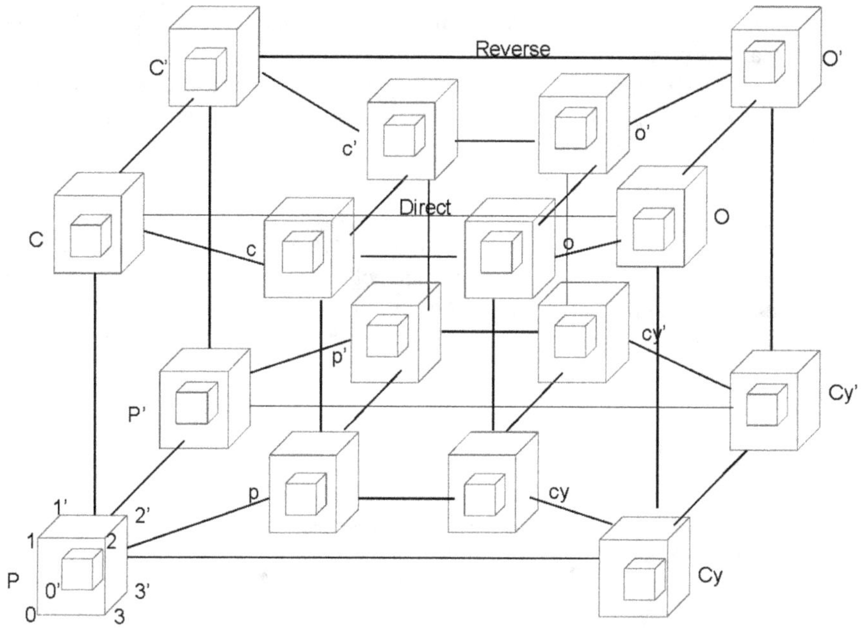

Fig. 7.17 Polytope 8D for command and control

For the polytope shown in Fig. 7.17 we selected a grafting strategy for modifications.

Any 0D component of 4D is replaced by a 4D component.

The notations are:

- Physical'-P',
- Cognitive'-C',
- Organizational'-O'
- Cyber'-Cy'

These denote the modified main stages

The modification may be operated for the stage or for the sub-stage.

The minuscule, p, c, o, cy and modified, p', c', o', cy' refer to the control frame.

The 8D Polytope as shown in Fig. 7.17 is a general framework for self-syncronization in command and control organizations (Liu et al 2011).

The physical network, cyberspace, cognitive network, and organizational network are integrated and overlapping. For instance, although the sensor network reports the targets' information, the information is then transferred by the communication network.

With the help of information services and situation analysis services, command and control teams, organized as a social network to match the objective, receive a common operational picture of the observed field. They can then define objectives and prioritize the effort matrix by recommending and evaluating courses of actions using scenario exploration, and computational experiments.

The CPSS may instruct and activate the enabler network to act. The CPSS maintains synchronization with the real world as well as self-synchronization of the human and physical system. The proposed framework couples humans and physical systems in observe, orient, decide, and act (OODA) loop. Human beings are included in the CPSS, not outside the system boundaries.

Figure 7.17 also emphasizes human-system integration in command and control. Human observation, orientation, decision, and action are supported by systems, and the systems make supplementary decisions and perform routine work according to the organizational intelligence of human beings.

References

Boyd, J.R.: Organic design for command and control. In: Science, Strategy and War-The Strategic Theory of John Boyd, pp. 189–200. Routledge (2007)

DeLaurentis, D.A., Crossley, W.A.: A taxonomy-based perspective for systems of systems design methods. In: IEEE International Conference on Systems, Man and Cybernetics, pp. 86–91 (2005).

DoDAF, The Department of Defense Architecture Framework, Version 2.0, vol. 01. Introduction, Overview & Concepts, Manager's guide (2009)

Dreier, A.S.: Strategy, Planning & Litigating to Win. Conatus Press, Boston, MA (2012)

Fischer, K.W.: Dynamic cycles of cognitive and brain development: measuring growth in mind, brain, and education. In: The Educated Brain: Essays in Neuroeducation, pp. 127–150 (2008)

Gustavsson, P.M., Planstedt, T.: The road towards multi-hypothesis intention simulation agents architecture - fractal information fusion modeling. In: 37th Winter Simulation Conference, pp. 2532–2541. ACM Press, Orlando, FL (2005)

Houdé, O.: 3-System Theory of the Cognitive Brain, A Post-Piagetian Approach to Cognitive Development. Routledge, London (2019)

Kahneman, D.: Thinking, Fast and Slow. Farrar, Straus and Giroux, New York (2011)

Liu, Z., Yang, D.S., Wen, D., Zhang, W.M., Mao, W.: Cyber-physical-social systems for command and control. IEEE Intell. Syst. **26**(4), 92–96 (2011)

Lupasco, S.: Le principe d'antagonisme et la logique de l'énergie. Prolégomènes à une science de la contradiction. Hermann & Co., Paris (1951)

Mahmood, A., Montagna, F.: A system of systems architecture framework (SoSAF) for production industries. In: 7th International Conference on System of Systems Engineering, pp. 543–548. Genoa, Italy (2012)

Marshall, S.P.: Schemes in Problem Solving. Cambridge University Press, Cambridge UK (1995)

Papanastasiou, M.B.: More than just plan, prepare, execute, and assess: enhancing the operations process by integrating the design and effects-based approaches. School of Advanced Military Studies United States, Kansas (2009)

Paris, C., Hall Johnston, J., Reeves, D.: A theoretical framework and measurement. Strategy for training team tactical decision making (1998)

Richards, C.: Certain to Win: The Strategy of John Boyd, Applied to Business. Xlibris, Philadephia, PA (2004)

Smith, E.A.: Complexity, Networking, and Effects-Based Operations: Approaching the "how to" of EBO. Boeing Company. Arlington, VA (2005)

Thiel, T.: The design of the connection machine. Des. Issues **10**, 1, 5–18 (1994)

Chapter 8
Outlook

Abstract The correlation of main levels of reality is presented using the MBIC (material, biologic, intelligent, cognitive) framework. An illustration of high dimensionality in reference architectures is presented. The diversifying and unifying methodology for science and engineering, for virtuality and reality is highlighted. The 8D Program manifesto, calls engineers, scientists, and entrepreneurs to identify and utilize observables, trends, mechanisms, and key issues concerning the emerging technology, industry and economy driving transition to higher dimension reference architectures and high complexity projects implementation.

8.1 MBIC

The convergence of several disciplines as, nano-science, biotechnology, information technology and cognitive science was studied under the label NBIC (Bainbridge and Roco 2006) or BINC (Rasmussen 2016).

Convergence was presented as a new paradigm that can yield critical advances in a broad array of sectors, from energy to health care, food, and climate (Sharp and Langer 2011).

It should be emphasized that the convergence does not quite grasp the essence of creativity required to explore high complexity domains. Tendencies to converge should be correlated with tendencies to diverge and it is the skillful correlation of both trends in a polytopic architecture that matters. The roads to higher complexity at all reality levels appear as an alternation between the convergence and divergence periods. This is significant for data analysis, equipments, technologies, industries and theories.

The proposed here MBIC, Material, Biologic, Intelligent, Cognitive, approach may be seen as an implementation of self-integrative closure concept were the convergence referring to matter in general is not restricted to nano domain (Iordache 2012, 2013, 2017). The idea for integrative closure is to reassess and completes the interrelation between the four basic ontological levels in the study of nature: material,

O. Iordache, *General Reference Architecture Frameworks*, Studies in Systems, Decision and Control 73, https://doi.org/10.1007/978-3-031-70718-6_8

biological, cognitive or psychological and intelligent or logical level (Hartmann 1952).

For MBIC concept the abbreviations correspond to Hartmann's four level of reality.

Information I, from NBIC or BINC concepts is replaced by intelligent I, from MBIC.

Moreover N for nano is replaced by M from material.

This corresponds to the recent change of focus from informatized to intelligentized frameworks. Obviously intelligence is based on information.

It was acknowledged that the hierarchical structures cannot serve as unique model for multi-level knowledge organization. Facing complexity the task of knowledge integration remains pertinent. Integrative closure is a strategy allowing exploring complexity.

The integrative closure aims to make ends meets, for the four levels or realms, emphasizing the hypothetical interconnection between the material realm, M and intelligent realm, I. The closed system is supposed to cross the gap between these farthest levels. This significant crossing corresponds also to the cyber-physical systems. The cyber part is associated to I while the physical part is associated to M.

Integrative closure hypothesis is looking for structural analogies and a common methodology for different domains as bio-inspired systems, knowledge organization, problem solving or technological developments.

The growth in complexity is not only about differentiation or divergence but also about integration or convergence and coordination. The pursuit of evolvability towards further step of self-evolvability for systems imposed the study of polytopic architecture associated to the roadmap shown in Fig. 8.1.

Figure 8.1 shows the Polytopic Roadmap for MBIC.

Following Hartmann ontology, for the polytope shown in Fig. 8.1 we could identify the levels of reality:

- Material-M
- Biologic-B
- Cognitive-C
- Intelligent-I

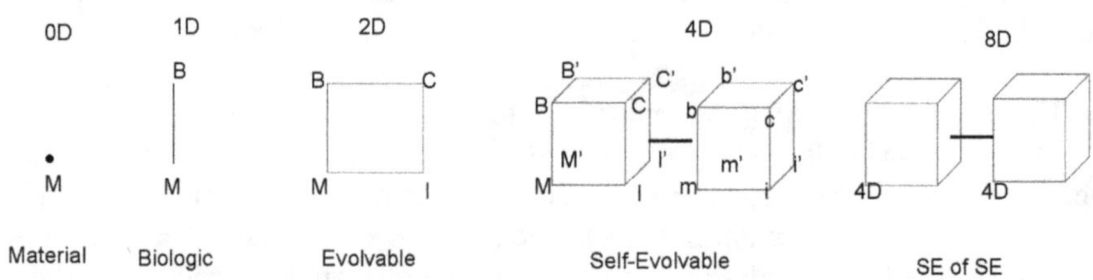

Fig. 8.1 Polytopic roadmap for MBIC

The basic levels are M, B, C and I have been represented on the front face of the outer cube as the Direct way.

Apostrophes ‘ mark the dual states.

M′, B′, C′ and I′ are represented on the back face of the outer cube of the polytope.

M’, B’, C’, I’ corresponds to the dual or Reverse way in the study of complexity.

Duality includes modifications.

The dimensions illustrated in Fig. 8.1 are:

- 0D-Material
- 1D-Biologic
- 2D-Evolvable
- 4D-Self-Evolvable
- 8D-SE of SE

We use the minuscule m, b, c, i to denote the control of Direct sequence and the minuscule m’, b’, c’, i’ to denote the control of Reverse sequence.

The 4D polytope shown in Fig. 8.1 illustrates the architecture of potentially self-evolvable systems. It is a representation consisting from a pair of 3D-cubes.

The second cube is a digital twin of the first cube.

The 4D-cube is obtained by joining all corners of the first 3D-cube with the corresponding corners of the second 3D-cube. The 8 lines connecting the two 3D-cubes are represented in Fig. 8.1 by a thick hyper-line. In this way the 4D-cube has a simplified representation. This representation is useful for Polytope 8D too (Thiel 1994; Ammon 1998).

The roadmap shows the 8D frame partially represented to avoid complicated drawings. In the 8D frame each 0D vertex of initial 4D are replaced by 4D for levels.

Figure 8.2 shows more details and a different presentation for the polytope 4D for MBIC presented in the Fig. 8.1.stage 4D (Iordache 2013, 2017).

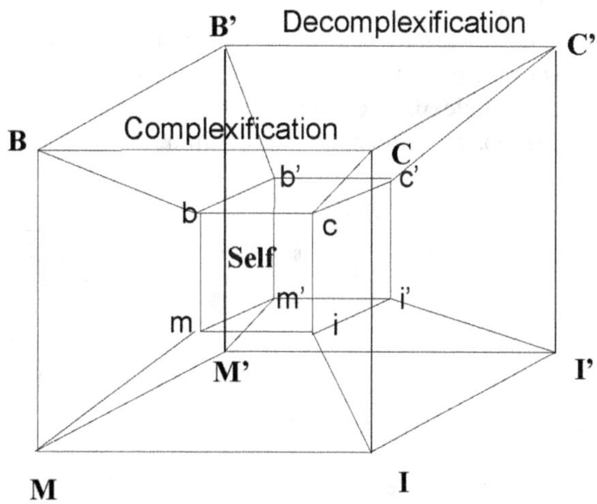

Fig. 8.2 Polytope 4D for MBIC

The polytope 4D outlines that in interaction with the converging, constructing or Direct way of Complexification M → B → C → I we need to look at the diverging, deconstructing or Reverse way of Decomplexification I′ → C′ → B′ → M′.

The swinging from the Direct way to the dual Reverse way is critical since the boundaries where new information is created consist of coexisting tendencies of unification and diversification.

Since dual complementary descriptions are intrinsically irreducible and in some cases contradictory, the concept integration or coordination of complementary descriptions requires a higher level description, a new dimension that emerges as a new hierarchical level of complexity. These are the 4D and beyond 4D reference architectures.

The inner cube, marked as the Self permits dialogue and interaction of the two ways in duality relation. These ways correspond to construction or Direct way of Complexification from M to I, or to deconstruction or Reverse way of Decomplexification from I′ to M′.

The Self needs to endorse the dialogue, to ensure coordination and coherence of evolution on different levels. Figure 8.2 shows that the structure of the inner cube is isomorphic to that of the outer cube. Observe that the two cubes of 4D-cube are reunited in a compact way. The 4D-cube may be considered as a two scale frame.

After any step in the Direct sequence of Complexification, the systems make contact with the Self represented by the inner cube. Signals as m, b, c and i reinforce the evolution according to the Direct sequence of Complexification. If the Self have to change the direction shifting to i’, c’, b’, m’ the next step will be along the Reverse sequence of Decomplexification. The set (m’, b’, c’, i’) corresponds to the control of Reverse sequence.

Figure 8.3 shows a developed roadmap for polytope 4D for MBIC.

The 2D and 4D frames are presented in a developed form here.

The 2D frames are based on Complexification. Complexification implies Decomplexification.

In the next step a new frame called the Self or Included Middle is considered.

The Self is shown with Complexification, c part and Decomplexification, d part. Since Complexification and Decomplexification frames are 2D the c and d frames are 2D too. They illustrate the conditions.

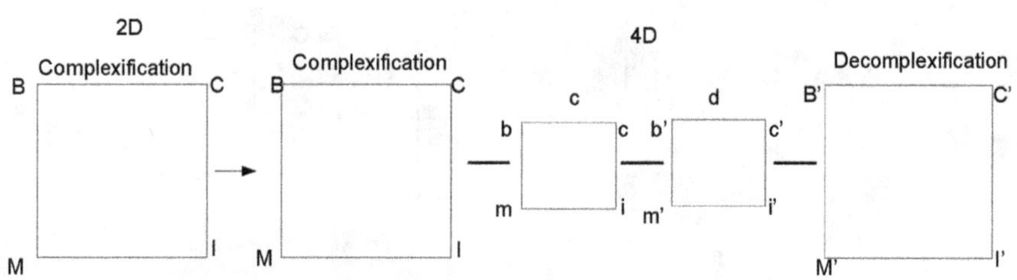

Fig. 8.3 Developed roadmap for polytope 4D for MBIC

Relating c and d allows obtaining an inner 3D-Cube not represented here. Finally the 4D frame results by inclusion the inner 3D frame in the outer 3D frame. It will be a 4D-cube representing the self-evolvable systems. The inner cube is a digital twin of the outer cube.

The developed roadmap may be interpreted in terms of dual process theory (Kahneman 2011). The Complexification M → B → C → I pertains to System 1 while the Decomplexification I′ → C′ → B′ → M′ pertains to System 2.

Lupasco (1951) and Houdé (2019) go beyond the approach based on System 1, (S1) and on System 2, (S2) to propose a theory in which the hidden third T, or the inhibitory-control system, System 3 (S3) takes precedence. T or S3 denotes the bridge between S1 and S2. S3 is associated to the Self and put together the c and d parts of IM.

The new presentation outlines the isomorphism between the reality levels and the virtual control systems.

The grafting strategy that maintains the 4D architecture but replace any 0D vertices by developed and structured 4D vertices is considered. It gives a hypercube of hypercubes that is a 4D of 4D. The new dimension is 8D = 4D + 4D. The first 4D come from the initial 4D frame as shown in Fig. 8.2 and the next 4D from the component 4D frames replacing the dots of the initial hypercube. Hypercubes as M refers to stages while hypercubes as m refers to control.

Figure 8.4 shows another presentation the polytope 8D for complexity.

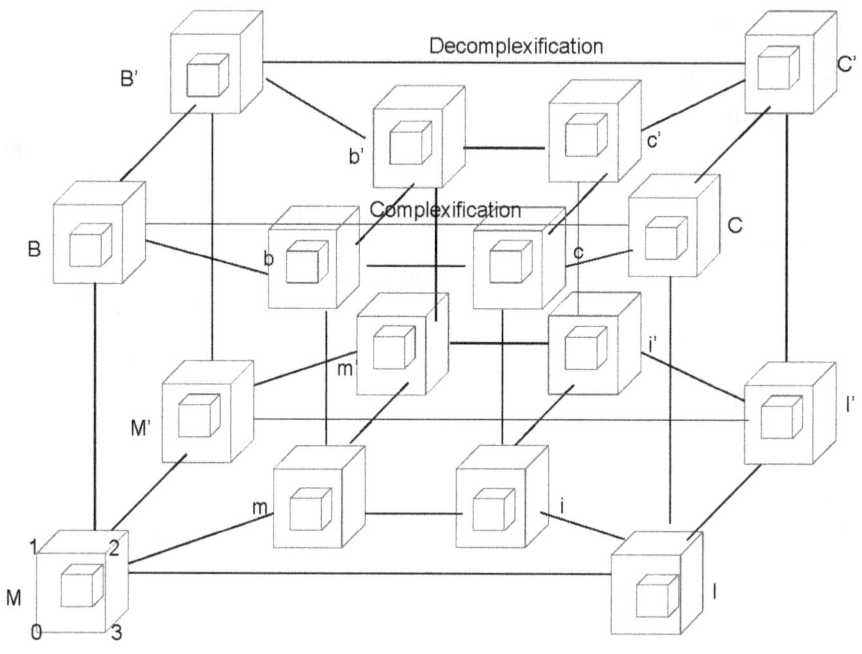

Fig. 8.4 Polytope 8D for MBIC

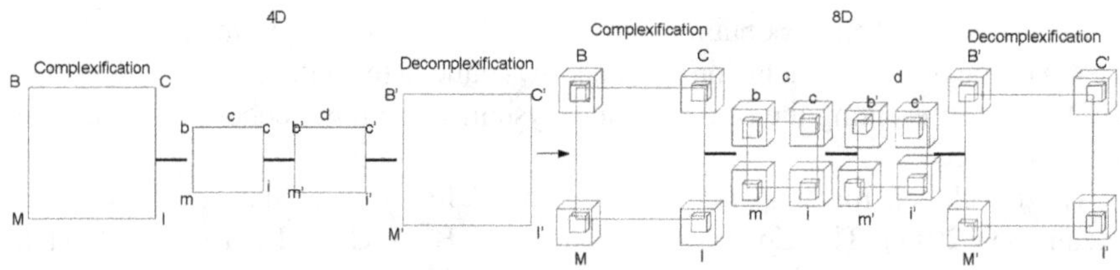

Fig. 8.5 Developed roadmap for polytope 8D for MBIC

This is the GRAF associated to MBIC.

The presented roadmaps outlined two sequences for evolution the direct sequence of Complexification M → B → C → I that corresponds to the developmental stages and the reverse sequence of Decomplexification M′ → B′ → C′ → I′ that corresponds to the dual modified stages. Apostrophes 'denote the dual or modified states.

For any reality level we may consider decompositions in sub-levels.

For instance for M we may consider, 0-micro, 1-nano, 2-meso, 3-macro. It is a multiscale approach as considered in chemical engineering. Another choice may be based on the basic type of forces: 0-electromagnetic, 1-gravity, 2-strong, 3 –weak. This approach was considered in physics.

The Polytope 8D is based on an intelligent type of development. The directly acquired development is questioned and modified by the reverse sequence. The bottom-up construction from direct sequence complements the top-down deconstruction of the reverse sequence.

Figure 8.5 shows a developed roadmap for polytope 8D for MBIC.

The 4D and 8D frames are presented here.

The transition from 4 to 8D is achieved by replacing any 0D vertex from the initial 4D architecture by component 4D frames. It is the grafting strategy that keeps the initial 4D architecture but replace any 0D vertices by component 4D vertices.

It results a hypercube of hypercubes. The new dimension is 8D = 4D + 4D. The first 4D come from the initial 4D frame while the next 4D from the component 4D frames replacing the 0D dots of the initial 4D hypercube

At this stage an inclusion of a new frame called the Included Middle or Self is considered. The Self is shown with Complexification c, part and Decomplexification d, part. Relating c and d allows obtaining the inner 7D-Ccube. Finally the 8D frame results, by inclusion of the inner 7D-cube in the outer 7D-cube. The inner 7D-cube is a digital twin of the outer 7D-cube.

Figure 8.6 shows the roads to higher dimensional MBIC modeling.

The dimensions of stages shown in Fig. 8.6 are:

- 0D-Material
- 1D-Biologic
- 2D-Evolvable
- 4D-Self-Evolvable
- 8D-SE of SE

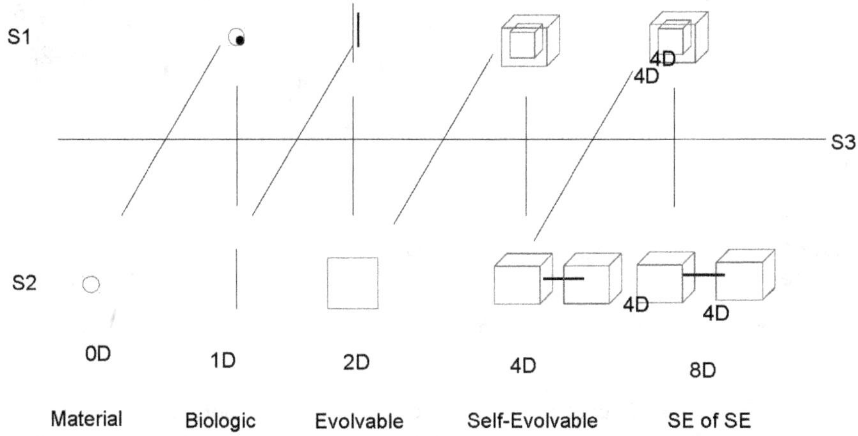

Fig. 8.6 Roads to higher dimensional MBIC

S1 and S2 roads have been outlined here. S2 is based on Aristotelian logic while S1 support a non-Aristotelian logic.

Consider that the S2 road starts from Material systems, 0D, represented here as a circle. Included in this an S1 frame appears. It is represented as an included dot. The dot and the circle will be related in a segment or line, 1D, as shown on the S2 road. The segment corresponds to Biologic systems.

Then on the S1 road an included middle represented as a segment appears. Transition to S2 road imposes the development of the segments. By relating the two segments the 2D square results. The square is associated to Evolvable systems.

Then on the S1 road an included cube appears and develops. Transition to S2 road imposes the development and the correlation of the two cubes. The 4D-cube is resulting.

It is associated to Self-Evolvable systems.

The 4D-cube on the S2 road is obtained by joining all corners of the first 3D-cube with the corresponding corners of the second 3D-cube. The 8 lines connecting the two 3D-cubes are represented in Fig. 8.6 by a thick hyper-line. In this way the 4D-cube has a simplified representation. Finally the 8D, SE of SE systems develop.

The 8D is resulting as 4D of 4D architecture ensuring intelligent-like capability.

The process may continue in the same way as a dance or a zigzag between S1 and S2 roads. Significant are the stages and the rhythm of performing these.

8.2 Endless Column

The trend toward 8D architectures is general and may be observed outside sciences and engineering.

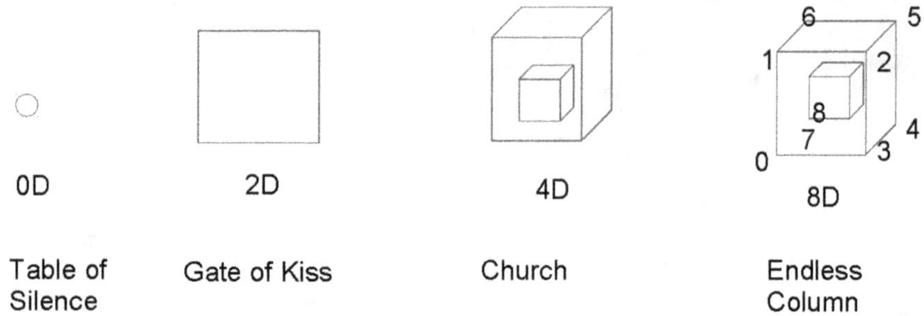

Fig. 8.7 Polytopic roadmap for avenue of heroes

Artistic accomplishments as for instance the Isenheim polyptich of Matthias Grünewald, the constructions due to Xenakis, or to von Spreckelsen, are inspiration sources for 4D-Art and 8D-Art.

The masterpiece of landscaping and sculpture that Constantin Brancusi left at Târgu Jiu, not far from the village of Hobita where he was born in Romania, is a significant expression of contemporary art. The site marks the spot where the Romanian soldiers heroically tried to stop the advancing invasion of their country in 1916.

The work comprises three parts: the Table of Silence, the Gate of the Kiss, and the Endless Column. They lie along a 1.7 km-long line that passes through the city, starting from the River Jiu and in line with the Church of Peter and Paul, a mediating point along the so-called Avenue of Heroes.

The Avenue of Heroes in Targu Jiu is described here as a polytopic roadmap in Fig. 8.7. Monuments on this Avenue are represented as little circle (Table of Silence), 2D rectangle (Gate of Kiss), 4D-frame (Church) and 8D-frame representing the Endless Column.

The dimensions illustrated in Fig. 8.7 are:

- 0D-Table of Silence
- 2D-Gate of Kiss
- 4D-Church
- 8D-Endless Column

Notice that there exist other identifications for the road elements on the Avenue.

For instance the Table of Silence may be associated to 1D. Endless Column is shown in a compact version in Fig. 8.7.

For 4D and 8D frames, developed architectural expressions have been implemented.

Arguments to associate 4D to the church follow Mithen (1966) approach to the prehistory of mind. According to Mithen primitive man have four kinds of intelligence that is four modules in the brain, evolving independently. The four modules are:

- Social intelligence: the ability to deal with other humans

- Natural-history intelligence: the ability to deal with the environment
- Tool-using intelligence
- Linguistic intelligence (language, metaphor and analogy)

The hunters-gatherers of pre-history were experts in all these domains, but those different kinds of expertise did not mix, that is are not correlated. The cognitive fluidity, when happens, caused the cultural explosion of art, technology, and religion. This corresponds to transition to 4D. Notice that the mind modularity and cognitive fluidity was comprehensively studied in psychology (Karmiloff-Smith 1996).

The cognitive fluidity supports the basic plan of churches with four related modules or areas. Another illustration of such developed 4D-frame is shown in Fig. 1.1. Figure 8.7 shows details of the 8D-frame. It is a closed form.

The 8D-frame shows two cubes with vertices 0–7 for external cube and 8–15 for internal cube.

Figure 8.8 shows the Polytope 8D for Endless Column.

The 16 elementary modules have been denoted from 0 to 15.

Figure 8.9 shows an open or developed form of the 8D-frame.

The 16 vertices of the 8D frame have been represented not in the closed compact form as in Fig. 8.8 but as an open developed vertical segment with 16 elements each corresponding to a 4D-frame in the original 8D-frame.

The Endless Column symbolizes the concept of infinity and the infinite sacrifice of the soldiers. The Endless Column stacks 15 octahedral modules, with a half-unit

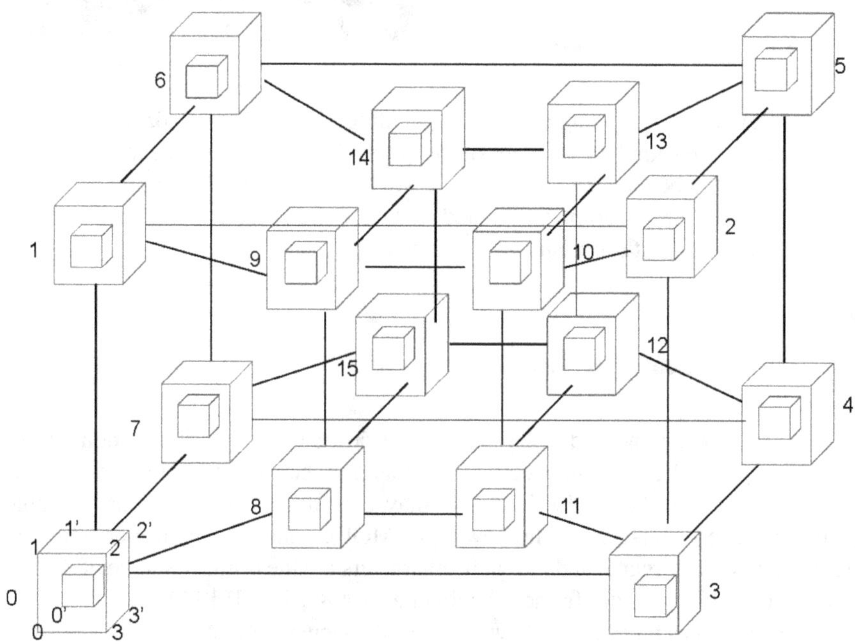

Fig. 8.8 Polytope 8D for endless column

Fig. 8.9 Endless column

at the top and bottom, making a total of 16 modules. The sub-levels 0, 1, 2, 3 and 0', 1', 2' and 3' correspond to the edges of the octahedral modules. The incomplete top unit is thought to be the element that expresses the concept of the infinite.

In the 1950s, the government planned to demolish the Endless Column and turn it into scrap metal, but this plan was never achieved.

8.3 Unifying and Diversifying

It is a persistent demand to develop roads and bridges between different domains of the same science, between sciences and engineering, between virtuality and reality.

For the case of chemical science we may consider the inorganic and organic chemistry. The metal–organic frameworks, (MOF), consist of both inorganic and organic units. The organic and inorganic characters are the primary source of duality and versatility for reticular frames (Yaghi et al. 1998, Li et al. 1999).

For the sciences unity example we refer to the General System Theory GST.

The thesis of the unity of science says that in principle sciences as physics, chemistry, biology or sociology must be part of a unified intellectual endeavor. The unity of science thesis is associated to the framework of levels of organization in nature, where physics is considered as the most basic level, chemistry the level below physics, biology below chemistry, sociology below biology, and so on. Further, cells, organisms, and societies are all biological, but they represent three different sub-levels of biological organization.

Sciences of matter, biosciences, cognitive sciences and intelligence sciences stand for the sciences classification based on Hartmann's ontology (Hartmann 1952; Salthe 2012).

The unity of sciences has been considered in terms of the circle of sciences. Logic is the foundation for mathematics, which is the foundation for physics, and physics is the foundation for chemistry, which is the foundation for biology, which is the foundation for sociology, the moral sciences, psychology, and the theory of knowledge, and this theory of knowledge is based on logic closing in this way the circle of sciences (Piaget 1967; Bailly 2010).

The unity of science thesis was tentatively argued for by several authors (von Bertalanffy 1962, 1968; Lupasco 1962). Developing the General System Theory, GST, Bertalanffy, stressed that the unity of science is granted, not by the reduction of all sciences to physics but by the structural uniformities of the different levels of reality. Nevertheless, by claiming the existence of similar laws that apply to all fields, at different levels, he was in fact claiming that reality has a logical structure. In his words, there is a structural correspondence or logical homology of systems. So, Bertalanffy endorsed the need for a logico-mathematical field, the subject matter of which is the formulation and deduction of those principles and laws which are valid for systems in general.

One inquiry is how is science used in engineering? Scientific and mathematical tools are used by engineers when creating, testing, and analyzing designs. Engineers construct physical and computer models that allow them to explore relationships between variables and recognize patterns in their data.

Science builds knowledge of how the natural world works, engineering uses that knowledge to develop useful technologies and these technologies may, in turn, provide observations and tools that help scientists build even more knowledge of the natural world. So science and engineering stand out as complementary.

For example, the scientific part of chemical engineering consists in breaking down complex systems which are then described using our understanding of fundamental phenomena. The engineering part of chemical engineering consists in using the gained knowledge, even if incomplete, in the design of a product which matches the desired characteristics (Wintermantel 1999). This is the relation between science and engineering and between virtuality and reality.

Polytopic Roadmaps are cognitive-intelligent inspired methods, useful for artifacts building, information representation, designs, operations and control. Polytopic Roadmaps as encountered at different levels of reality and in different artificial applications domains have been presented in order to emphasize the implemented isomorphism of the knowledge frameworks and of control mechanisms (Iordache 2021).

The main objective of Polytopic Roadmaps is to understand and make use of development stage theories as suggestions for new artificial operations, equipments, technologies, industries and societies. Polytopic Roadmaps are tools for thought and for organizing high complexity projects implementation.

The re-application of this thought in developing new ways of study or explanations of material-bio-cognitive-intelligent relevance for reality systems development should be considered as unavoidable.

Let S1 and S2 denote System 1 (thinking fast) and Systems 2 (thinking slow) as studied in dual processes theory, DPT (Kahneman 2011).

In terms of DPT, unifying of engineering and reality is close to S1 systems while diversifying sciences and virtuality is close related to S2 systems.

S1 is an attempt to impose artificiality, to build the isomorphism and the unifying framework for different technologies industries, economies and so on.

S2, try to find concepts, laws, and models in various fields, to provide a diversified framework for different sciences of nature or society.

The difference between S1 and S2 is the difference between the engineering approach, and the sciences approach, between reality and virtuality. These approaches should work together in duality and unity.

The search for S1 and S2 is based on the idea of construction by iteration of frames of frames over frames and so on. The processes of different levels will show specific properties. This means that we have to consider iterations in new conditions and not just recursive procedures. So, we need to associate iterations to inclusion.

One objective is to highlight the elements of the hypothetical unified engineering for S1, or the hypothetical unified science for S2.

To explore complexity we need both S1 and S2. The roads to higher complexity at all reality levels appear as a zigzag between the self-reinforcing information on the S1 road and the expansiveness of the S2 road.

The proposed here Polytopic Roadmaps for organizational concepts can be considered as tools for scientists building models and for engineers for the evolution or realization of high complexity systems. In this sense we consider the self-evolvable of self-evolvable architectures not at the level of a theory, but as frameworks for a theory of theories.

The Polytopic Roadmaps are based on the equation: Method: Iteration + Inclusion.

This is based on Lupasco principle of antagonistic dualism and the principle of included third.

The applicability of the proposed approach has to be proven in specifying the general concepts and translating them into the languages of peculiar scientific disciplines or of peculiar engineering projects. The objective is to outlines roads and bridges between virtuality and reality. We start from the observed parallelism or isomorphism of reference architectures for different domains. The logical reference architecture of our virtual conceptual schemes is such that it corresponds to the structural reference architecture of reality, natural or artificial. Parmenides quote summarizes such observations of universality:

> For it is the same thing that can be thought and that can be.

Fields ranging from inorganic materials to biosystems and cognitive systems, to design and logico-mathematical models, socio-economic systems share similar reference architecture frameworks.

For example, correlating the physical with cyber in cyber-physical systems, CPS, allows developing industry 4.0 and everything 4.0 that is 4D live-like systems.

Then, grafting 4D for any 0D component of the previously developed 4D live-like system shows the way to 8D intelligent-like systems.

A repeated finding of this book is that apparently self-evolvability observed in natural systems is based on the same directed sequence of reference architectures as the self-evolvability of concepts in human mind and of the associated engineering projects.

Two explanation of this result of universality may be imagined (Winiwarter 1986a, b, 1999). Either we can describe reality because the organizational principles of our concepts are isomorphic to the organizational principles in physical, chemical, biological, cognitive, and intelligent systems, or we describe all real phenomena according to the organizational principles of our concepts.

In the first explanation, human scientific mind is only a special form of self-evolvable of self-evolvable reality on the recently emerged level of intelligence.

In the second explanation the origins of the universal organizational principles of our mind remain to be explained.

Preliminary answers are offered by the Logic of Contradiction or Logic in Reality (Lupasco 1952, 1962; Brenner 2008).The principle of antagonistic dualism and the principle of included third of Lupasco show that between the two explanations, of objectivity and of subjectivity, it is a common sense, universal place for both.

8.4 The 8D Program Manifesto

- **Sources**

Engineers, researchers, and entrepreneurs encounter in their activity high dimensional reference frameworks naturally associated to high complexity projects. Problem solving for complexity domains require polytopic projects and new methodologies. Polytope is the general term of the sequence: point, segment, polygon, polyhedron and so on.

The main source for polytopic projects is the multiplicity of processes, scales, fields, dimensions, disciplines and logic rules for the considered systems. Multiple interactions and emergences are the core of higher complexity and of associated models and projects. Multiplicity is responsible for the label "poly" in polytopic projects. Studies of complexity should be considered as resources for polytopic projects.

We refer to chemical and biological complex systems, artificial life, artificial brains, artificial intelligence, and to complex mathematical systems.

Gardiner et al. (2010) use EEG data as evidence of higher dimensional fractal signals in the brain, ranging fractionally from 5 to 8D. This suggests that the brain operates in a higher than 4D frame of space–time. So, the 8D may be considered as a safe choice for intelligent-like capabilities.

A significant source for polytopic projects are the studies in Blue Brain Project, BBP, showing that brain's connections exist in 8D or higher dimensions. The 4D and 8D frames proved to be significant in the neurodynamics of brain waves, for data processing.

A recent source for polytopic projects is the advent of industry 4.0, of society 4.0, and of everything 4.0 developments.

The 8D is of interest for algebraic topology and algebraic combinatorics (Loday and Vallette 2012). Self-evolvability SE, is necessary for 4D live-like (A-life) as for 8D intelligent-like (A-intelligence) capabilities. For 8D we refer also to SE of SE.

Hypercubes 4D allowing live-like capability as that existing in industry 4.0 are the building bricks of the proposed 8D architecture associated to intelligent-like capability.

Standing on the achieved 4D frames, the 4D of 4D that is the 8D frames represent the new challenge. A quantum leap forward from 4 to 8D, that is from life-like to intelligent-like capability is required. This is achieved by grafting 4D on the primary 4D framework.

- **Methods**

Encoding similar structures in different branches of sciences and engineering, of virtuality and reality, the 8D Program looks for methodological unification.

The Method to explore high complexity is based on a successive application of diversifying and unifying stages. The Method is abbreviated by the equation: Method: Iteration + Included Middle.

Iteration corresponds to diversifying while Included Middle corresponds to unifying step.

Iteration is accompanied by antagonistic duality (Lupasco 1951).

Let S1 and S2 denote System 1 (thinking fast, non-standard) and Systems 2 (thinking slow, standard) as studied in psychology (Kahneman 2011).

Any system, S2, is accompanied by the dual non-system, S1 and reversely. Iteration and relation allows the evolution from the standard system S2 to the new system S1S2.

The Method may be applied locally or globally on the reference architecture. This corresponds to step by step or to grafting strategies.

Some of the considered pairs of Systems S1, S2 and their applications are presented in the Table 8.1.

Category theory was considered by much of the mathematical community as abstract "non-sense". Opposed to this, the set theory represents the common "sense".

Table 8.1 Synthesis of systems applications

Systems (S2) standard	Non-systems (S1) non-standard	Complexity roots	ReferencesIordache	Domain
Markovian	Non-Markovian	Multiple processes Multiple levels	(2010)	Polystochastic
Archimedean	Non-Archimedean	Multiple fields Multiple scales	(2011)	Complexity
Cartesian	Non-Cartesian	Multiple dimensions Multiple closures	(2012, 2013)	Polytopic Higher dimension
Aristotelian	Non-Aristotelian	Multiple logics Multiple disciplines	(2022)	Higher dimension Transdisciplinarity
"Sense" sets	"Non-sense" categories	Multiple theories Multiple technologies		diversify unify Virtuality-reality

Labeling an argument as "abstract nonsense", is not intended to be critical.

Method assumes that both type of thinking systems, S2 and S1 are necessary and that evolution implies the inclusion of third system S3 (identified as included middle, IM, hidden third T, or Self). S3 or T describing the interaction of S1 and S2, defines the bridge between S1 and S2 (Lupasco 1951; Houdé, 2019) and it is committed to creativity features.

The frame S1TS2 will characterize the post-standard evolution.

So, we can propose roadmaps as: states, processes, S1, S1TS2 and so on that is: 0D, 1D, 2D, 4D and so on.

There exist roadmaps to 4D (self-evolvable, live-like) and to 8D (SE of SE, intelligent-like). Rules of T make the resulting architecture to be live-like, and then intelligent-like.

Figure 8.10 illustrates the building steps of the proposed General Polytopic Roadmap, GPTR: 0D, 1D, 2D, 4D and 8D. As complexity increases, the dimension increases.

Notations are: K0, K1, K2, K3 for hierarchy of conditions.

We use K0', K1', K2', K3' for reverse sequence.

We use the minuscule to denote the control sequences.

The dimensions illustrated in Fig. 8.10 are:

- 0D-Descriptive
- 1D-Adaptive

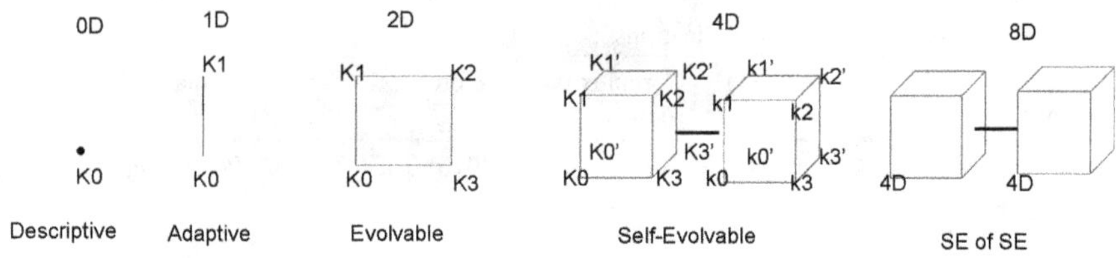

Fig. 8.10 General polytopic roadmaps

- 2D-Evolvable
- 4D-Self-Evolvable
- 8D-SE of SE

The 4D polytope shown in Fig. 8.10 illustrates the architecture of potentially self-evolvable systems. It is a representation consisting from a pair of 3D-cubes. The 4D-cube is obtained by joining all corners of the first 3D-cube with the corresponding corners of the second 3D-cube. The second cube is a digital twin of the first cube.

The 8 lines connecting the two 3D-cubes are represented by a thick hyper-line. This representation is useful for Polytope 8D too.

Figure 8.11 shows details of the Polytope 8D. It is a 4D of 4D frame.

As above, we use majuscule for real modules and minuscule to denote the control sequences.

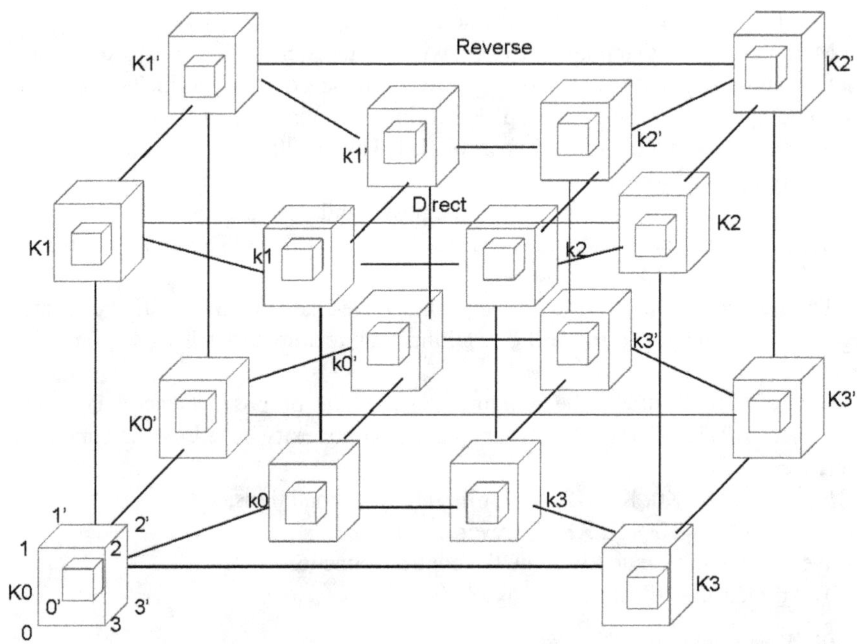

Fig. 8.11 Polytope 8D

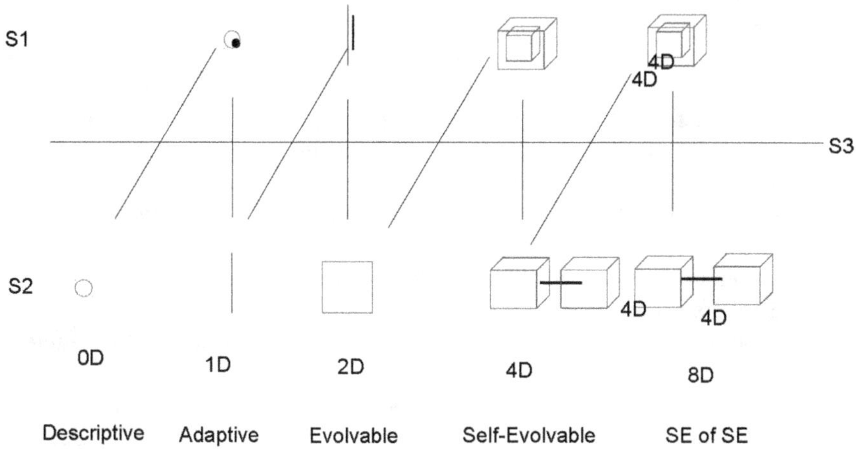

Fig. 8.12 Roads to higher dimensional general polytopic roadmaps

The sub-levels in K0 may be denoted by 0, 1, 2, 3 for a direct sub-sequence in K0, and by 3', 2', 1', 0' for a reverse sub-sequence in K0. For a complete description of K0, the sub-self of K0 components should be identified too. This will complete the 8D.

Figure 8.12 shows the roads to higher dimensional General Polytopic Roadmaps.

S1 assumes a non-standard logic. S2 is based on standard logic. S3 describes the interaction of S1 and S2. The S2 road starts from 0D frame Descriptive represented here as a circle. Included in this an IM frame develops. It is represented as a dot. The circle and the dot have been related in a segment, 1D corresponding to Adaptive frame.

Then on the S1 road an included segment appears and develops. Transition to S2 road imposes the development of the two segments. By relating the segments the 2D square result. It corresponds to Evolvable systems. Then on the S1 road, a square of squares organized as a cube, then including another cube appears. It is the 4D frame associated to Self-Evolvable systems.

The 4D-cube on the S2 road is obtained by joining all corners of the first 3D-cube with the corresponding corners of the second 3D-cube. The 8 lines connecting the two 3D-cubes are represented in Fig. 8.12 by a thick hyper-line. In this way the 4D-cube has a simplified representation. This is useful for 8D frame too.

In the next step the SE of SE systems develop by 4D grafting to primary 4D stage. The 8D is resulting as 4D of 4D architecture ensuring intelligent-like capability. The process may continue indefinitely in the same way as a dance or a zigzag between S1 and S2 roads. Significant are not only the stages but also the rhythm of performing these stages.

Observe that the GPTR from Fig. 8.10 includes stages from S2 roads.

Other GPTR of interest may start with S2 stages and continues with S1 stages.

This corresponds to a change of focus from sciences to engineering that is from S2 to S1. This is based on the observation that in this century, we are on the cusp of an epoch making transition, from being passive observers of Nature to being active designers of Nature (Kaku 1997). The era now allows us to harvest the fruits of the last century of Science. The Age of Discovery in Science is coming to a temporary slow-down, opening up an Age of Engineering. To explore complexity we need a specific zigzag between S1 and S2, between engineering and sciences, between reality and virtuality.

It is an iteration process and an attempt to solve high complexity problems by reference architectures dimensionality increasing and by inclusions.

The Table 8.2 illustrates the main steps of polytopic roadmaps for different domains of knowledge.

Notations are: SM-supramolecular, MOF-metal–organic framework, PP post-principles, SE-self-evolvable, TD-transdisciplinarity, SO-self-organized, SoSoS-S^3-systems of systems of systems, R-real, C-complex, H-quaternions, O-octonions, MBE-Model Based Engineering.

- **Developments**

The Method aims to highlight and to take advantage of the isomorphism or parallelism between roadmaps for different reality levels.

This isomorphism is the root of the 8D reference architecture, GRAF, resulting as 4D of 4D. This means that any 0D vertex of the 4D frame is replaced by a new 4D structure. Grafting to all vertices in the primary 4D frame allows the quantum leap forward to 8D frame.

Mathematical tools concerns iterating and interacting, doubling, contracting, relating, and grafting.

Material systems shown in Table 8.2 refer to SM and MOF. Then SM of SMs, MOF of MOFs, MOF on MOFs and MOP (Metal–organic polytopes) (Kondinski et al. 2022) follow.

Biosystems refer to Dynamical Hierarchies, Q-species, and Genetic Code. For Artificial Life mechanisms the 4D are Vesicles of Vesicles (Iordache 2022).

Cognitive systems refer to Development Stages for Piagetian and neo-Piagetian studies, due to Kurt Fischer, and other authors.

Table 8.2 show the example of Numbers and Normed Division Algebras: R, C, H, and O.

Octonions, O, are 8D. Another mathematical example refers to Leibniz-Loday algebras: associative algebras, dendriform dialgebras, quadri-algebras, and octo-algebras (Leroux 2003; Ebrahimi-Fard and Guo 2005; Loday and Vallette 2012).

Mathematical tools refer also to Combinatorial Species, Formal Concept Analysis FCA, Digital Twins, DT, Hopf Algebras, HA, and Operads (Iordache 2012, 2021).

Engineers aim to develop operations, equipments, technologies, industries, in successive stages as people develop from sensory-motor to formal, and then to post-formal stages.

Table 8.2 Synthesis of polytopic roadmaps

Dimension→ ↓ Domain	0D	1D	2D	4D	8D
Chemistry	Molecules	Polymers	SM	SM on SM	MOF of MOF SM-MOF
Chemicalengineering	Chemistry	Unit operations	Productengineering	Sustainableindustry 4.0	Intelligentindustry 8.0
Dynamical hierarchy	Molecules	Polymers	Micelles V-Vesicles	V on V = V2	V2 of V2
Brain waves		Gamma	Beta	Alpha	Theta
Developmentstages piaget	Sensori-motor	Pre-operational	Concrete + formal	Post-formal PF	PF of PF
Developmentstages fischer	Sensori-motor	Representation	Abstraction + principles	Post-principles	PP of PP
Complexity	Descriptive	Adaptive	Evolvable	Self-evolvable	SE of SE
Computing telephones	Manual	Desktop	Laptop	Smartphone tablet	Intelliphone polytope
Construction	Communal	Private	City regulated	Private/City	Intelligent homes, cities
Contradiction	States	Processes	Events, E	ETE': self-evolvable	SE of SE
Cybernetics	Cyber-less	Feedback	Reflexive predictive	Embodied SE	SE of SE
Defense	Human	Weapons	Machines	Robots drones	SO robots SO drones
Disciplines	Reality	Mono-disciplinary	Multi-disciplinary	TD trans-disciplinary	TD of TD
Economy	Traditional	Agriculture	Industrial	Service digital	Intelligent Collaborative
Education	Tradition	Memory	Open access	Knowledge production	Individualized SO
Knowledge	Tradition	Craft, art geometry	Sciencesengineering	Artificial-life	Artificial-Intelligence
Manufacture	Manual	Mechanic	Electric electronic	Digital self-organized	SO of SO Intelligent

(continued)

Table 8.2 (continued)

Dimension→ ↓ Domain	0D	1D	2D	4D	8D
Reliability	Practice	Specifications	Adaptivepredictive	Self-evolvable	SE of SE Intelligent
Society age	Stone	Metals	Industrial	Society 4.0: Soc 4.0	Soc 4.0 of Soc 4.0: Soc 8.0
Solving problems	Natural	Mechanism	Organism evolving	Self-evolvable	SE of SE
Systems	Objects	Processes	Systems S	SoSoS = S^3	S^3 of S^3
Division algebras		R Real	C Complex	H Quaternion	O Octonions
Modeling	Data	Model	Simulation	Digital-twin DT	DT of DT MBE
V-model	Data	W Waterfall	V Validation	V-model VM	VM of VM

As elementary basic operations we can consider partitions, permutations, classifications, graphs, complexifications and the corresponding iterations towards higher dimensions.

For equipments we need to evaluate multi-scales, distributors-collectors, cyclic-operations, middle-vessels, mixing, assembly and hybrids. Technology advances refer to polytopic flowsheets, polytopic OLAP (on line analytic processing), and polytopic engineering designs (Iordache 2021, 2022).

- **Perspectives**

The 8D column in Table 8.2 is about the research in progress and perspectives.

As perspectives we are looking to artificiality for main levels of reality (material, biologic, cognitive, and intelligent).

This refers to:

- Intelligent materials: SM-MOF, MOP, Quantum Computer, Compounds Coding
- Artificial life-A-Life, Bio-Molecular Computer, Genome Coding
- Artificial brains-BBP (blue brain project), Neural Coding
- Artificial intelligence-A-Intelligence, Intelligence Coding
- Artificial mathematics-ATP (Automated Theorem Proving) Artificial science

Unifying and diversifying for reality and virtuality represents the main project.

The future leading technological and scientific projects require coding exploration and target artificiality for materials, living, cognitive and intelligent systems.

Coding research starts from the periodic table of chemical elements and continues with Human Genome project and the BRAIN initiative (Jorgenson et al. 2015).

Living technologies, artificial life (Szostak et al. 2001; von Kiedrowski et al. 2010), artificial brain, artificial intelligence and smart systems are leading scientific and technological projects.

Living technology is promising because it shares the fundamental properties of living systems (Bedau et al. 2010). These include self-assembly, self-organization, growth and division, purposeful action, adaptive complexity, evolution, and intelligence.

Existing technologies become increasingly live-like and intelligent-like. Examples of living and intelligent technology projects are synthetic biology attempts to make living systems from scratch in the laboratory, systems exhibiting collective and swarm intelligence distributed across the world wide web, self-reconfiguring robots and drones, and others.

Advances in computer science go pointing the way to an intelligent future—one of autonomous self-evolvable experimentation, research and modeling.

In the new approach, artificial intelligence techniques are employed to carry out the entire cycle of scientific experimentation, including the origination of hypotheses and modeling to explain observations, the devising of experiments to test these hypotheses and models and the physical implementation of the experiments using robots to falsify hypotheses and models.

One development we might expect to see in this area over the next years is the construction of self-evolvable artificial scientists (Muggleton and Zauner 2006).

This would involve the confluence of active learning and autonomous experimentation systems with specific technology.

Autonomous experimentation and modeling will play an important role in meeting the challenge of accumulating and analyzing the comprehensive data sets that are outside normal affordable resource constraints including time and human bandwidth.

Moreover, such computational techniques capable of deciding from past observations which test to perform next are also of great interest for instrumentation with limited communication bandwidth where decisions need to be taken, as to which is the best next step to take, such as which experiment to perform, when to perform it or when to stop collecting data.

Mobile robots or drones exploring harsh environments such as deep sea locations typically are able to communicate back only a small fraction of the data gathered by their sensors. Accordingly, the robot or the drone itself would be in the best position to decide the next experimental and modeling step.

In the coming years, we can expect a confluence of wireless networks and lab-on-chip sensor technology with large-scale applications in environmental monitoring. Autonomous self-evolvable experimentation and modeling will be required to fully exploit the potential of the technology. In such a lab-on-chip network, each sensor node will be endowed with a limited supply of wet chemistry and accordingly can perform only a limited number of experiments.

Autonomous experimentation and modeling enabled by self-evolvable artificial scientists and automated theorems proving methods have begun to open up novel approaches (Bai et al. 2022, 2024).

Relating and unifying mathematical theories through topos-theoretic bridges has been reconsidered by Caramello (2016, 2017).

Complex industry needs polytopic industry and organization, Industry 8.0, RAMI 8.0, as next objectives. Complex society needs society 8.0, polytopic ecology, health, defense, and postmodernism.

- **Outline**

The 8D Program calls engineers, scientists, entrepreneurs and artists to proceed to identify, document and utilize observables, trends, mechanisms, laws and key issues concerning the emerging technology, industry and economy driven transition to higher dimension reference architectures. This means dimension 8D, as for instance industry 8.0, society 8.0, 8D-art and so on. Based on generalized 4D the advance to 8D became feasible, in principle, and should be considered.

It is a call for GPTR and GRAF that is for 8D polytopic projects implementation for all the reality levels: material, biologic, cognitive, and intelligent. The superposition of the four reality levels in a quintessential or human level should be considered too.

It is a call for live-like (4D) and for intelligent-like (8D) systems a call to develop both A-Life and A-Intelligence. It is a new Method for exploration and discovery based on Iteration and Included Middle, and on selecting the appropriate rhythms of diversification and unification.

A major feature is the perspectives offered by the hybridization and cross-fertilization of polytopic projects research due to their location on the cycle of sciences (material, biologic, cognitive and intelligence domains).

Study of the 8D and of the GPTR and GRAF may be our gateway forward into moving past the standards models to what lies before and behind the scenes, to the non-standard models, and after unifying the standard and non-standard, to post-standard into 4D and then to 4D of 4D that is to 8D.

The 8D Program outlines the possible evolution steps of physical, industrial, ecological and social understanding and projects implementation.

Introducing this ground-breaking field of most important practical and theoretical interest and the key areas for the higher steps for live-like and intelligent-like technologies, industries, and societies, the GPTR, the GRAF and the 8D Program will be of highest interest for our future.

References

Ammon, J.: Hypercube Connectivity within ccNUMA Architectures. SGI Origin Team, Mountain View, CA (1998)

Bai, J., Cao, L., Mosbach, S., Akroyd, J., Lapkin, A.A., Kraft, M.: From Platform to Knowledge Graph: Evolution of Laboratory Automation. JACS Au **2**, 292–309 (2022)

Bai, J., Mosbach, S., Taylor, C.J., Dogoncan, K., Lee, K.F., Simon, D.R., Akroyd, J., Lapkin, A.A., Kraft, M.: A dynamic knowledge graph approach to distributed self-driving laboratories. Nat. Commun. **15**, 462 (2024)

Bailly, F.: L'anneau des disciplines. Enquête sur quelques concepts théoriques et gnoséologigues. AFSCET, Paris (2010)

Bainbridge, W.S., Roco, M.C.: Managing Nano-Bio-Info-Cogno Innovations: Converging Technologies in Society. Springer, Berlin (2006)

Bedau, M.A., McCaskill, J.S., Packard, N.H., Rasmussen, S.: Living technology: exploiting life's principles in technology. Artif. Life **16**(1), 89–97 (2010)

Brenner, J.E.: Logic in Reality. Springer Science & Business Media, Berlin (2008)

Caramello, O.: The Theory of Topos-Theoretic 'Bridges'-A Conceptual Introduction. Glass Bead (2016)

Caramello, O.: Theories, Sites, Toposes: Relating and studying mathematical theories through topos-theoretic 'bridges', Oxford University Press (2017)

Ebrahimi-Fard, K., Guo, L.: On products and duality of binary, quadratic, regular operads. J. Pure Appl. Algebra **200**, 293–317 (2005)

Fischer, K.W.: Dynamic cycles of cognitive and brain development: measuring growth in mind, brain, and education. In: The Educated Brain: Essays in Neuroeducation, pp. 127–150 (2008)

Gardiner, J., Overall, R., Marc, J.: The fractal nature of the brain: EEG data suggests that the brain functions as a "quantum computer" in 5–8 dimensions. Neuroquantology **8**, 137–141 (2010)

Hartmann, N.: The New Ways of Ontology. Greenwood Press, Westport (1952)

Houdé, O.: 3-System Theory of the Cognitive Brain, A Post-Piagetian Approach to Cognitive Development. Routledge, London (2019)

Iordache, O.: Polystochastic Models for Complexity. Springer, Berlin (2010)

Iordache, O.: Modeling Multi-Level Systems Springer, Berlin (2011)

Iordache, O.: Self-Evolvable Systems. Machine Learning in Social Media Springer, Berlin (2012)

Iordache, O.: Polytope Projects. Taylor & Francis, CRC Press, Boca Raton (2013)

Iordache, O.: Implementing Polytope Projects for Smart Systems. Springer, Cham (2017)

Iordache, O.: Polytopic Roadmaps. Springer, Cham (2021)
Iordache O.: Roads to Higher Dimensional Polytopic Projects-Reference Architectures. Springer, Cham (2022)
Jorgenson, L.A., Newsome, W.T., Anderson, D.J., Bargmann, C.I., Brown, E.N., Deisseroth, K., Donoghue, J.P., Hudson, K.L., Ling, G.S., MacLeish, P.R., Marder, E.: The BRAIN initiative: developing technology to catalyze neuroscience discovery. Philos. Trans. R. Soc. B **370**(1668), 20140164 (2015)
Kahneman, D.: Thinking, Fast and Slow. Farrar, Straus and Giroux, New York (2011)
Kaku, M.: Visions: How Science Will Revolutionize the 21st Century. Anchor Books, Doubleday, New York (1997)
Karmiloff-Smith, A.: Beyond Modularity: A Developmental Perspective on Cognitive Science. MIT Press, Cambridge, MA (1996)
Kondinski, A., Menon, A., Nurkowski, D., Farazi, F., Mosbach, S., Akroyd, J., Kraft, M.: Automated rational design of metal−organic polyhedra. J. Am. Chem. Soc. **144**, 11713–11728 (2022)
Leroux, P.: On some remarkable operads constructed from Baxter operators (2003). arxiv:math. QA/0311214
Loday, J.-L., Vallette, B.: Algebraic Operads. Springer, Heidelberg (2012)
Lupasco, S.: Le principe d'antagonisme et la logique de l'énergie. Prolégomènes à une science de la contradiction. Hermann & Co., Paris (1951)
Lupasco, S.: Le principe d'antagonisme et la logique de l'énergie. Rev. Métaphysique Et De Morale **57**, 467–468 (1952)
Lupasco, S.: L'Energie et la matière vivante. Antagonisme constructeur et logique de l'hétérogène. Julliard, Paris (1962)
Mithen, S.: The Prehistory of the Mind: The Cognitive Origins of Art, Religion and Science. Thames and Hudson, New York (1966)
Muggleton, S., Zauner, K.-P.: Artificial scientists. In: Emmott, S., Rison, S. (eds.) Towards 2020 Science, pp. 36–37. Microsoft Research, Cambridge (2006)
Piaget, J.: Classification des sciences et principaux courants épistémologiques contemporains. In: Piaget, J. (ed.) Logique et connaissance scientifique, pp. 1151–1224, Gallimard, Paris (1967)
Rasmussen, S.: The "4th industrial revolution" in fact a more radical "BINC Revolution"? In: Conference American Chamber of Commerce, China, Beijing, 14 December 2016 (2016)
Salthe, S.N.: Hierarchical structures. Axiomathes **22**, 355–383 (2012)
Sharp, P.A., Langer, R.: Promoting convergence in biomedical science. Science **333**, 527 (2011)
Szostak, J.W., Bartel, D.P., Luisi, P.L.: Synthesizing life. Nature **409**, 387–390 (2001)
Thiel, T.: The design of the connection machine. Des. Issues **10**, 1, 5–18 (1994)
von Bertalanffy: General system theory - a critical review. Gen. Syst. **7**, 1–20 (1962)
von Bertalanffy, L.: General Systems Theory: Foundations, Development, Applications. George Braziller, New York (1968)
von Kiedrowski, G., Otto, S., Herdewijn, P.: Welcome home, systems chemists! J. Syst. Chem. **1**(1), 1 (2010)
Winiwarter, P.: Autognosis: The theory of hierarchical self-image building systems. In: Proceedings of the International Conference on Mental Images, Values & Reality. Society for General Systems Research (1986a)
Winiwarter, P.: Concepts of self-organization, self-organization of concepts. In: Proceedings of the International Conference on Mental Images, Values & Reality. Society for General Systems Research (1986b)
Winiwarter, P.: A periodic system of systems concepts. Semiotica **125**(1/3), 47–62 (1999)
Wintermantel, K.: Process and product engineering: achievements, present and future challenges: process and product development. Chem. Eng. Res. Des. **77**(3), 175–188 (1999)

Index

O. Iordache, *General Reference Architecture Frameworks*, Studies in Systems, Decision and Control 73, https://doi.org/10.1007/978-3-031-70718-6

T

U

V

Z

Printed in the USA
CPSIA information can be obtained
at www.ICGtesting.com
CBHW061841131024
15794CB00003B/41